The Ark of Millions of Years

New Discoveries and Light on the Creation

by

E.J. Clark & B. Alexander Agnew, PhD

Cover art by Holly Ford

authorHOUSE™

1663 LIBERTY DRIVE, SUITE 200
BLOOMINGTON, INDIANA 47403
(800) 839-8640
WWW.AUTHORHOUSE.COM

First published by AuthorHouse 07/11/05

ISBN: 1-4184-5928-3 (e)
ISBN: 1-4184-3402-7 (sc)
ISBN: 1-4184-3403-5 (dj)

Library of Congress Control Number: 2004092145

Printed in the United States of America
Bloomington, Indiana

This book is printed on acid-free paper.

The Dedication

To my beloved daughter, Eve, who was destined to ask, "The question," and to my beautiful and faithful cat, Youshabel, who patiently laid by my side during the writing of this book.

-E.J.Clark

To my family, who over the years has been kept awake by my dreams. Together we have run our fingers over the seams of the universe and marveled at the tailor's workmanship. I have learned that the word is the key to the natural high. It emits pictures from those who cannot paint; music from those who cannot play. With it, the power of the mind is infinite. May they never fall asleep.

-B. Alexander Agnew

Acknowledgments

Every effort has been made to contact the owners and publishers of copyrighted material, photos, and illustrations used in this book. Any errors or omissions are purely unintentional and will be corrected as soon as possible. Some photographs and illustrations are by the authors, Dover, and uncopyrighted internet sources. Quotation sources are given in the footnoted bibliography at the end of each chapter. Other sources for illustration and photos are cited in the text, and some are noted below.

Reprint of photo of Giant Stucco Mask from pyramid K-5 at Piedras Negras, c. 1933, used with permission from the "University of Pennsylvania Museum (neg. #000000)."

Chinese characters found in *The Union of the Polarity* are excerpted from *The Discovery of Genesis* ©1979 Concordia Publishing House. Reproduced with permission under license number 04-5-17.

Table of Contents

The Foreword

It all began when my daughter, Eve (and wouldn't you know she had to be named Eve), a university student, called home one evening to vent her frustration over a comment made by one of her professors. He said, "You don't believe that old story of Noah's ark do you?" Of course that statement got the attention of everyone in his class. He continued, "Why that ark would have to be the size of the state of California to hold all the animals."

My daughter then wanted to know how the ark managed to hold all the animals. I told her that I didn't know the answer at present but I would seek the answer. For sixteen years thereafter I was diligent in my search for the evasive answer to her question. Then about one and a half years ago the answer slowly came, in bits and pieces, over a two- month period of time by inspiration. The inspired revelation nearly "knocked my socks off" and to coin an old adage, " was way beyond my wildest imagination."

I reasoned if such an event truly occurred then there must be some documentation somewhere preserving an account of this incredible story. After searching through many ancient writings and scriptures, to my amazement, the story was preserved as I had received it, in bits and pieces, a little bit here and a little bit there. When compiled, the bits and pieces fell together like a jigsaw puzzle telling a new story of the creation of the earth, the fall of Adam and Eve, Noah and his ark, the flood, and the destiny of this world. The story that has emerged will forever change your perspective on the creation, no matter what your religious persuasion might be.

What I didn't realize, in the beginning, was that I had accidentally stumbled onto ancient knowledge that had been lost for **6,000** years! This book restores the lost ancient knowledge that is symbolized by the Star of David or Seal of Solomon pictured on the front cover. Knowledge that answers so many questions that here to fore have been major stumbling blocks. Stumbling blocks that have pitted science against religion, and creationists against evolutionists.

My daughter has since asked, "How do you know that this is true? Could one writing have influenced another?"

I replied, "Well yes, it could have but it is all that we have to work from. Furthermore, since the story came first by inspiration and lastly by documentation, I feel that it is true." I want to emphasize that I do not claim to be a prophet, seer, or revelator, but rather a seeker of knowledge. Knowledge can be available to all that earnestly seek it. Do not the scriptures in the Holy *Bible* teach this principle in the following verses?

"Verily, Verily, I say unto you, Whatsoever ye shall ask the Father in my name, he will give you......ask, and ye shall receive." (*John* 16:23-24)

Desiring that all should know this compelling story, I found a co-author to help me write this tremendous work. I call it a tremendous work because in order to tell the story we had to write about the entire creation, no small feat. The answer to my daughter's question is given somewhere in this book. Sorry, there are no giveaways here. You must read the book in order to find the answer.

This book is non-denominational. It is not about religion but about the creation; however, it is difficult to separate the two because the story was preserved in ancient and modern sacred scriptures and writings. We do not aspire to teach religious doctrine of any kind but only to present what we found through documented sources.

The documentation tells the story. We merely guide it. A word of caution; read each chapter in sequence, don't skip around and risk becoming lost and confused. Although the story is profound and deep, it is written in simple terms so that all may understand what is conveyed. Whenever possible we have used the exact text as it is worded.

E.J. Clark

I have a curiosity. Not an ordinary curiosity. No, not by any means. At thirteen I stated that every thing we do will eventually affect the universe. One day I was reading about Aristotle creating the terms and definitions for government. At least aristocracy, democracy, anarchy, and monarchy were described by him and are used today as textbook definitions of government.

But, the most remarkable thing about the story was how he did it. "All knowledge is innate," he wrote. He felt that if a man just sat and thought things through hard enough he could derive all knowledge from whole cloth. The difference between people that were awake, and those were asleep became profound. To act and to be acted upon were a stupor of thought to the masses. Einstein is often quoted as saying that imagination is more important than intelligence.

Well, I was well endowed with imagination and have since my youth refined my ability to bring those images into reality either through teaching, writing, or actually building the darned things with my own hands. Wondering how things are made, or can be made, is to me the most exciting exercise in the human experience. Feeling authorized by some of the greatest thinkers of all recorded history to think about things, and formulate theories, I endeavored to go beyond the furthest point of man's exploration.

It has been my great fortune to be born at the dawn of the modern information age. The ability to find and consume knowledge is unfettered. The knowledge imparted to students by even the most efficient and accomplished professor—assuming they are in class and paying attention—is a ten point dot on a football field compared to the information one can assimilate with time and the hunger to make it known.

Now you know my source of energy. When I met E.J. Clark I was presented with a most marvelous challenge. Assist in the compilation and analysis of the creation of the earth utilizing the most modern physics and ancient observations available. At first glance it was a strange dichotomy. But, after a few late-night forays into dozens of amazing transitory relationships, the truth opened up to both of us.

The ancient texts were without any doubt astounding in their fascinating ability to cross-check and testify of cosmic and terrestrial events. I doubt there are six people on the planet who have at their disposal a better working knowledge of these texts than E.J. Clark. Further, without the physics in hand one could never understand the observations of the prophets and other ancient writers. Without the ancient texts, the modern scientist is unable to correlate the creation without gaping holes in logic.

Putting the two collections together, we created this landmark literary work. For the first time in over 6,000 years the creation of the earth has been explained from two simultaneous viewpoints like binoculars whose visionary sum will inspire common man and bring him very close to understanding one of the great mysteries of heaven. Who made the earth and placed it here may or may not be important to you. How it was made and how it was transported here is monumentally important. How it has been transformed and will yet be transfigured is enough knowledge to keep the pure scientist and the pure theologian intellectually and spiritually aroused for the rest of mortality and beyond.

Sit back and be prepared to have all sciences and religions poured into your head. Relish your satisfaction as volumes of pure truth assume their proper position in the libraries of your mind. Then marvel as the great observatory of your imagination opens to the stars and reveals where you came from and why you are here. One of the real goals of this book is to convince you that you are eternal being having a mortal experience and that your potential is absolutely unlimited. Where you go from here is completely up to you.

To quote a famous intergalactic traveler's favorite launching slogan, "To infinite and beyond." (Buzz Lightyear—Pixar, 1998 *Toy Story*)

B. Alexander Agnew, PhD

NOTE: There are numerous illustrations and pictures throughout our book. Many of these are far more effective when seen in color. You may go to www.arkofmillionsofyears.com and view or print these for free. If you have any questions or comments, you may email us at info@arkofmillionsofyears.com .

The Beginning

The Ark of Millions of Years contains 21 chapters starting with the first chapter, titled ***The Beginning*** and ending with the last chapter, titled ***The End.*** In Hebrew the word for beginning is chakhmah which denotes wisdom and the word for end is binah which means understanding. Therefore this book begins with and imparts wisdom and ends with understanding.

The beginning of wisdom starts with this chapter. We have made a bibliography of some of the books, historians, documents, and terms, we have referenced that the average reader may not be familiar, listed in alphabetical order to be used as a reference chapter. We recommend that you briefly scan over this chapter and in the following chapters if you are not familiar with a word, name, or title, just refer back to this chapter for a brief description of that term.

1. **Apocrypha**, a collection of books or parts of books belonging to the *Old* and *New Testaments*, including writings or statements of

doubtful authorship or authority. Some of these books or parts of books belonging to the *Old Testament* in the Vulgate (authorized *Bible* of the Roman Catholic Church) are not admitted to versions in Protestant use.

2. **Atarhasis text**, is an Assyro-Babylonian poem compiled in Greek by the Babylonian priest, Berosus, written no later than 1700 B.C. This text is one of the three surviving Mesopotamian documents to give an account of the flood. Khisutros, Xisitros, Ziusudra, and Atarhasis are the different names given to the same mythological person Utnapishtim, identified as Noah, who is to be found in a Sumerian text.

3. **Babylonian Talmud,** is the body of Jewish civil and canonical law. It is considered the oldest of the various Talmuds in current use. The Talmud is a repository of thousands of years of Jewish wisdom. The oral law, which is as ancient and significant as the written law (*Torah*), is included. The complete printed edition of the Talmud was produced by Daniel Bomberg (a Christian) in 1520-30.

4. **Bahir, the Book of Illumination**, is one of the oldest and most important of all classical Kabbalah texts. It remained the most influential and widely quoted primary source of ancient Kabbalah until the publication of The Zohar. The head of the Safed School of Kabbalah, Rabbi Moshe Cordevero (1522-1570), says, "The

words of this text are bright (Bahir) and sparkling, but their brilliance can blind the eye." First printed edition appeared in Amsterdam in 1651.

5. **Berosus** was a third century B.C. Greek Babylonian/Chaldean historian and priest of the temple of Bel.

6. **Book of Abraham,** is a document written upon papyrus by Abraham, in his own hand while he was in Egypt. The ancient writing was found in the catacombs of Egypt and eventually fell into the hands of Joseph Smith, the Mormon Prophet, who translated the record. The papyrus record was destroyed in the Chicago fire. The text is part of Latter Day Revelation called the Pearl Of Great Price, sacred scriptures used by the Church of Jesus Christ of Latter-Day Saints.

7. **Book of Enoch The Prophet,** is translated from the original Ethiopian Coptic script, written in the second or first century B.C., first translated by Richard Laurence in 1821 and was published in successive editions, ending in the 1883 version. The early church suppressed the writing and it was believed to be lost until its discovery in 1773, by Scottish explorer James Bruce in Ethiopia.

8. **Book of The Secrets of Enoch,** was written in Egypt by a Hellenistic Jew. The original Greek text has been lost but the Slavonic text was revealed to the world in 1892. For 1200 years

this book was known only to a few people in Russia. This book was also referred to by Origen and used by the Church Father, Irenaeus. W.R. Morfill translated the Slavonic into English shortly after the text was found.

9. **Ferrar Fenton's _The Holy Bible in Modern English,_** was first published in 1903. Fenton was a linguist fluent in the Greek, Hebrew, and Chaldean languages. He studied _The Bible_ in its original languages to ascertain what its writers actually said and taught. Having mastered these languages, he translated the _Old_ and _New Testaments_ directly into English, with footnotes. His translation has been called the most accurate rendering of the Scriptures. Over the years, his translation has been in and out of print and at one time was difficult to find. It is now back in print.

10. *The Book of Jasher* also called **The Book of the Ancient Word** and **The Book of the Upright,** was translated into English from the Hebrew in 1840. It is referred to in Joshua 10:13 and Second Samuel 1:18. The 1840 text is the only authentic record. The Jasher account supplies many details not given in *The Bible*, from Adam down to the time of the Elders, who immediately succeeded Joshua.

11. *Book of Mormon,* is an ancient record that extends from 600 B.C. to 421 A.D. In or about the latter year, Moroni, the last

of the Nephite historians, sealed the sacred record and hid them in a stone box in the ground. In 1827, the same Moroni, then a resurrected personage, delivered the engraved plates to Joseph Smith who translated them. The book is a record of a remnant of the house of Israel who was led by the hand of God to build a ship and cross the waters to the promised land, present day Mexico. First English Edition published in 1830. The Latter Day Revelation is part of the sacred scriptures used by the Church of Jesus Christ of Latter- Day Saints.

12. **Book of Moses,** was revealed to Joseph Smith the Prophet, in June 1830. It is part of Latter Day Revelation contained in the *Pearl of Great Price*, sacred scriptures used by the Church of Jesus Christ of Latter- Day Saints.

13. **Dead Sea Scrolls,** were discovered in caves along the shores of the Dead Sea in the late forties and early fifties. The Qumran texts and fragments have been translated as precisely as possible and fifty key documents were published in *The Dead Sea Scrolls Uncovered*, translated and interpreted by Robert H. Eisenman and Michael Wise, published in 1992. Later, more scrolls were translated by Wise, A. Begg, Jr. , and Cook. They were published in *The Dead Sea Scrolls; A New Translation* in 1996.

14. **Doctrine and Covenants (D&C),** is the only book in existence, which bears the honor of a preface given by the Lord himself.

It is modern revelation dictated by Jesus Christ and revealed through the prophets of the Church of Jesus Christ of Latter-Day Saints. It is not a book just for the Latter-day Saints but intended for the whole world for their salvation. It is part of the sacred scriptures used by the Church of Jesus Christ of Latter –Day Saints.

15. **Egyptian Book of The Dead** or **The Book of the Great Awakening** as it is otherwise known, is the papyrus of Ani, an Egyptian text written in hieroglyphics and is probably the Theban version which was much used from the XV111th to the XXth dynasty. The papyrus was found in an 18th Dynasty tomb near Luxor in a perfect state of preservation in 1888. It was a copy of the Egyptian Book of the Dead, written around 1500 B.C. for Ani, Royal Scribe of Thebes, Overseer of the Granaries of the Lords of Abydos, and Scribe of the Offerings of the Lords of Thebes. It was typical of the funeral book in vogue among the Theban nobles of his time. Originally published in 1895, by the order of the Trustees of the British Museum.

16. **Gilgamish Epic,** is one of the three Mesopotamian-Sumerian-Chaldean accounts of the flood. Gilgamish was the fifth ruler in the Dynasty of Erech, which was considered the 2nd Dynasty to reign after the Deluge. In 1854-5, 12 tablets were discovered in the ruins of the Library of Nebo and the Royal Library of Ashur-

bani-pal at Nineveh, now in the British Museum. The tablets revealed an epic story of Gilgamish searching for immortality and an account of the story of the Deluge. George Smith translated and published the accounts in 1873.

17. **Hebrew Book of the Dead: In the Wilderness,** is an ancient text translated with added commentary by Zhenya Senyak in 2003. It is a spiritual book fusing together Biblical text and some key concepts of Kabbalah. This text was more than a thousand years old when David was King in Jerusalem.

18. **Herodotus** was an ancient Greek historian (484?-424? B.C.). His history was the first to be written in an organized, chronological manner that earned him the title "Father of History." He is famous for nine books that he wrote on the rise of the Persian Empire, the Persian invasions of Greece in 490 and 480 B.C., the heroic fight of the Greeks against the invaders, and the final Greek victory. The things he learned in his travels formed the materials of his histories.

19. *The Holy Bible,* is a library of sixty-four books for Protestants, seventy-three for Roman Catholics, and twenty-four for Jews. The word *Bible* means books. These books are the sacred writings of Judaism and Christianity. Some of the authors of these books are known, others known by tradition, and some are unknown. The older part of *The Bible*, called the *Old Testament*, contains

the sacred writings of the Jews. It is also a record of the history of the Jews and contains the code of laws by which they were governed. The *New Testament* gave the Christian revelation of God to "all the world," speaking "to the Jew first, and also the Gentile." The Jews today accept only the *Old Testament*, but the Christians believe that the two parts have equal authority. The first printed edition of *The Bible*, by Johannes Gutenberg, was in 1456. *The Bible* has remained the world's best seller for many years. No other book can equal its record.

20. **Josephus** was born 37 A.D., four years after the crucifixion of Jesus, in Jerusalem, the son of a priestly aristocratic Jewish family on both sides. He served as a priest and later became the commander of Jewish forces in Galilee following the revolt against Rome that began 66 A.D. He spent his later life in Rome under the patronage of the Roman emperors, who considered him a prophet, where he composed his history of the Jewish people and his account of the Jewish war that led to the destruction of Jerusalem and the Temple in 70 A.D. No source, other than the *Bible*, provides more relevant information on the first century than this historian. He died ca. 100 A.D. His writings have been published in *The New Complete Works of Josephus* in 1999.

21. **Kabbalah** means traditions and may be defined as esoteric Jewish doctrine. Kabbalah is divided, in general, into three categories,

theoretical, meditative and the magical. The principle doctrines of the text are designed to solve the following problems; those of creation, the Supreme Being, the cosmogony, creation of angels and man, the destiny of man and angels, the nature of the soul, the nature of angels, demons, and elementals, the mysteries contained in the Hebrew letters, the transcendental symbolism of numerals, and the import of the revealed law.

22. **Kebra Nagast** or **The Glory of the Kings,** is the oldest of all the Ethiopian Scriptures. It was translated by Sir E.A. Wallis Budge and published in English in 1932, under the title of *The Queen of Sheba and Her Only Son Menyelek.*

23. **Nag Hammadi Library,** is the only complete one-volume, modern language version of the renowned library of fourth-century papyrus manuscripts discovered in Egypt in 1945. Because the library's tractates derive from the Hellenistic sects now called gnostic, and survive in Coptic translations, it is characterized as the Coptic Gnostic Library. The original manuscripts of the library are fragmentary in many places, but improvements in translation and new tractate introductions have made for a more complete translation. *The Nag Hammadi Library in English* was published in 1990.

24. **Popol Vuh,** meaning **Book of Community,** is the Sacred Book of the ancient Quiche Maya. During the middle of the sixteenth

century a literate Quiche Indian transcribed his folk legends into Latin script. In the late seventeenth century Father Francisco Ximenez, parish priest of Santo Tomas Chichicastenango, copied and translated the now lost manuscript. The book is a mixed record of the cosmic beliefs, folklore, semi-historical migrations and genealogies of the Quiche Indians, one of the Mayan tribes that lives in the highlands of Guatemala. Printed in English in 1950.

25. **Pseudepigrapha,** are those scriptures which have been eliminated by various councils in order to make up the standard *Bible*. They were usually eliminated due to historical inaccuracy but nevertheless retain historical value because some truths are contained in the writings.

26. *Qur'an,* is the sacred scripture of Islam. Moslems believe that the Angel Gabriel revealed the *Qur'an* to Mohammed a little at a time from about 610 to 632 A.D. The name *Qur'an* means *the reading* or *that which is to be read.* Mohammed, gave the name of *Qur'an* to these messages when he had received only a part of them. He called his teachings *Islam.* After he died, his followers collected his revelations and made them into what is now called the *Qur'an.* It consists of verses grouped into 114 chapters, or *suras.* The suras vary in length from only a few lines to many verses. The *Qur'an* stresses above all that there is

only one God, whom the Moslems call *Allah*. Its teachings and authority form the basis of the great Islamic civilization.

27. **Sepher Rezial Hemelach** or ***The Book of the Angel Reziel***, also known as ***The Divine Book of Wisdom*** and ***Adams Book***. This book was translated, in 2000 into English, from the ancient Hebrew in the rare and complete 1701 Amsterdam edition by Steve Savedow. The back cover of the book reads, "According to Hebrew legend, the *Sepher Rezial* was presented to Adam in the Garden of Eden, given by the hand of God, and delivered by the angel Rezial."[1]

28. **Sefer Yetzirah** or **The Book of Creation,** is without question the oldest and most mysterious of all Kabbalistic texts. It is so ancient that its origins are lost to historians but traditionally it had been attributed as the work of the Patriarch Abraham. The revised edition was published in England in 1977.

29. *Torah* is the Jewish name for the first five books of *The Holy Bible*, *The Pentateuch*, or *"Law of Moses."* In Jewish literature it could mean a law or precept and even mean Divine instruction as in revelation.

30. **Vedic Texts** are Hindu sacred texts and writings. *The Rig Veda* is the earliest of religious text of Hinduism. The gods of Hinduism are many and varied. The Vedas are organized into four sections: 1. *Samhita,* hymns to the various elements

and deities 2. *Brahmana,* collection of prose that describes various ritualistic details 3. *Aranyaka,* texts which deals with philosophical concerns 4. *Upanishads,* contain the highest form of philosophical introspection. The *Bhagavad Gita,* a portion of the *epic* poem *Mahabharata,* expresses the Hindu belief that the spirit is subject to an indefinite series of existences or reincarnation. *The Ramayana,* a counterpart to the *Bhagavad Gita,* details the exploits of Rama.

31. **When the Earth Nearly Died,** book written by D.S. Allan & J.B. Delair (two British scientist), published by GateWay Books, Bath, UK., in 1995. Their book presents the compelling evidence of a catastrophic world change 9,500 B.C., when all life on Earth was almost wiped out by appalling conflagration, overwhelming flood, and global catastrophe caused by a cosmic interloper. It is a serious scientific study of the event. Republished under the title, *Cataclysm!,* by Allan & Delair. [2]

32. **Zoroaster, Zarathustra** is the founder of Zoroastrianism, one of the historically great religions, and practically unknown in the world today outside India. The religion can be described as belief in one God (monotheism) who gives guidance and direction to his people through laws and commands (ethics). It was the state religion of three great Iranian empires, which flourished from the sixth century B.C. to the seventh century

A.D. The surviving sacred books of the religion are known as the *Zend Avesta*. They comprise 1. *The Yasna*, containing the Gathas, the songs or hymns which are ascribed to Zarathustra 2. *The Yashts*, sacrificial hymns to various deities 3. *The Vendidad*, the law against demons dealing mainly with ritual impurity. The later *Avesta* books are mostly concerned with the end of time.

33. **The Zohar** or **The Book of Splendor,** is the fundamental book of Jewish Kabbalism. It is a work of great antiquity attributed to be the work of the ancient Sage, Simeon ben Yohai. However scholars of *The Zohar*, generally believe that it never could have sprung from the brain of one single man. It is a compilation of a mass of material drawn from many layers of Jewish and non-Jewish mystical thought, covering many centuries. Its teachings are found in the oldest portions of the Babylonian and Palestinian Talmuds as well as in Jewish Apocalyptic literature produced in the centuries right before and after the destruction of the second Temple. The Jews regard it as sacred literature. *The Book of the Veiled Mystery, The Secrets of the Torah,* and *The True Shepherd*, are books contained within *The Zohar.* The first printed editions appeared in 1588-90. Today the teachings of *The Zohar* are contained in five volumes.

"The value of Wisdom cannot be measured, nor

understanding of Knowledge"

The Sefer Rezial Hemelach or *The Book of the Angel*

Rezial[3]

End Notes

[1] *Sepher Rezial Hemelach: The Book of the Angel Rezial, edited and translated by Steve Savedow with permission of Red Wheel/Weiser, York Beach, ME and Boston, MA. To order call 1-800-423-7087*

[2] *Cataclysm!: Compelling Evidence of a Cosmic Catastrophe in 9500 B.C.* by D.S. Allan & J.B. Delair, Bear & Co. a division of Inner Traditions International, Rochester, VT 05767. Copyright ©1995 & 1997 by D.S. Allan & J.B. Delair.

[3] et al *Sepher Rezial Hemelach:*

The Dimensions

It is important for the reader to understand that not only is there clear evidence in ancient records of the existence of multiple heavens—what may be described in more scientific terms as dimensions—and that there is evidence of travels between these heavens. Whether it be a beam of light, a sound, a particle, or a planet there is mounting evidence that these dimensions exist.

Though it may seem blasphemous to some, research into the trans-dimensional universe has uncovered some remarkable things. Theories describing the relationship between matter and energy are beyond most folks, but we will attempt to put some of the ideas and information in a form the reader can understand. Einstein and his associates uncovered a portion of the energy stored in the nucleus of a single proton. His equations inspired scientists and mathematicians to blend theories of time travel with the revelation of multiple dimensions. Black holes spinning so fast that they do not collapse, but rather possess a calm center through which one might view or even pass into another universe. The proof that our universe is indeed expanding and rotating at the same time, coupled with abundant negative energy, also provides a sound denominator for equations charting a path across the universe in a single step. He also put into universal language the basic relationship between matter and energy. He described a space-time continuum in which time and distance are not linear at all.

In the early 19th century a simple farm boy named Joseph Smith, Jr. in upstate New York revealed the same things in revelations he recorded. The irrefutable fact is that matter and energy, or light or spirit as some may say, are interchangeable. Although man seems to understand this relationship, no one has revealed the process. Changing from energy into light and into matter is not a repeatable experiment. Even changing from matter into energy has only been done to a partial degree. We can burn wood, produce electricity with chemicals, and even split atoms apart. But, we cannot get the little neutrons and protons to actually convert into 100% pure energy. Man has been thoroughly astonished by the energy contained in a few pounds of atoms.

The space-time continuum casually woven through science fiction is real. Think of it as a bed sheet pulled tight at every edge. It is flat and free of wrinkles. That is unless something distorts it. No matter how taut the fabric, placing a bowling ball in the center of the sheet will distort that plane. Any matter moving through space will cause a similar bend or a warp in the space-time continuum. The straight line that once was the thread of the sheet's fabric now curves around the bowling ball. The same thing happens with the continuum around the earth, the sun, and any other body of matter that moves through space. A gravitational wave is produced each time mass moves through space. Airplanes make a small and measurable distortion. Planets make a large distortion, which can be detected with special observatories.

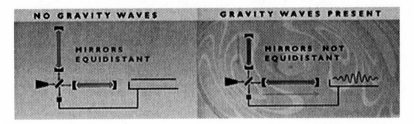

LIGO is the Laser Interferometer Gravitational Wave Observatory. There are several of them now working. They are the culmination of experiments designed and performed by scientists since the early 1900's. To detect gravitational waves, LIGO's two L-shaped detectors—one in Livingston, Louisiana, the other in Hanford, Washington—work in concert across the nearly 2,000 miles (3,000 kilometers) that separate them.

Gravitational radiation causes a strain on the fabric of space-time transverse to (i.e., extending across) the direction in which its waves are propagated. As they strike, say, the Earth, the waves will stretch the fabric in one direction, while along another, they compress it.

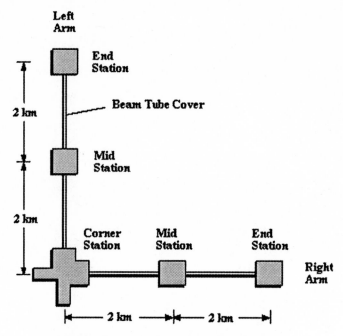

Schematic layout of LIGO Site at Hanford, WA
(Installation at Livingston, LA has no mid-stations)

The distances between the test masses—fitted with fused silica mirrors and hung in enormous vacuum pipes encased in concrete—will be measured with a laser beam split in two, each half then traveling the length of each arm multiple times. By then recombining (or interfering) the separate beams, any change in the distance between either of the pairs of masses will throw them out of phase with each other, thus indicating—if all goes well—the form of the gravitational wave as it reaches the Earth. If no waves pass, the distances between the test masses will be the same. To make sure, the two detectors are located far, far apart. That way, no local interference -- earthquakes, noise or fluctuations in the lasers -- can be mistaken for a gravitational wave. [1]

This means that in reality there is a shorter path between two points than the straight line of the thread. There is a hypotenuse that may significantly reduce that distance. But, in order to utilize that shorter path, one must leave the bed sheet universe and access a different dimension. Travel across that gap would likely not be detectible in the *known* universe. That does not mean it is not there.

Even in *The Holy Bible*, there are clear references to transdimensional travel by mortal beings and by vessels.

"And when they were come up out of the water, the Spirit of the Lord caught away Philip, that the eunuch saw him no more: and he went on his way rejoicing. But Philip was found at Azotus and passing through he preached in all the cities, til he came to Caesarea...And *Peter* saw heaven opened, and a vessel descended unto him, as it had been a great sheet knit at the four corners, and let down to the earth...And this was done thrice: and the vessel was received up again into heaven."[2]

Many modern physicists are familiar with the works of Helmoltz, Lord Kelvin, Faraday, Maxwell, and other giants whose constants and theories form the foundations of the science. They laid the groundwork over 100 years ago for a highly controversial discovery that three-dimensional physics is only a subset of *hyperspatial* dimensions. The evidence being gathered by latter-day instruments indicate these other dimensions not only control the physics of our planet, but the stars and possibly even life itself.

21

Direct observations made with instruments and human senses show that information transfer between dimensions is occurring. The community forming right now has formulated the idea that this transfer of information, and possibly even physical matter, is the foundation substrate for everything in this dimension...the one from which we are writing and you are reading.

The mathematical parameters for this transfer—often called *gating*—were founded in the work of a few 19th Century pioneers. These giants not only challenged the bounds of contemporary math, they wrote new math and developed new constants that to this day have proven to be sound. German mathematician George Riemann, Scottish physicists Sir William Thompson and James Clerk Maxwell, and British mathematician Sir William Rowan Hamilton proposed a radical explanation for the basic properties of solid matter. They developed the *vortex atom*.

Most 19th Century scientists believed one of many theories about matter from the Roman poet Lucretius' *hard bodies*—endorsed by Newton as well—to other atom-like models with fixed shapes and electron orbits. *Vortex atoms* were envisioned as tiny whirlpools of moving matter in an *aether*—an incompressible fluid that made up space in the universe.

William Thomson, the young professor, aged 22

In 1873, Maxwell successfully explained the math that proved the existence of still smaller particles known as *quaternions.* These four complex numbers—it's easier to explain particles these small by mathematical properties than actual appearance—were readily adopted by Hamilton. These mathematical rules explained how light and energy move in the form of waves throughout the universe. In 1897, A.S. Hathaway stated that instead of three dimensions there were in fact at least four in a groundbreaking paper, "Quaternions as numbers of four-dimensional space," published in the *Bulletin of the American Mathematical Society.*

James Clerk Maxwell as a young man

Maxwell, a poet as well as a mathematician, was already sure that this three-dimensional physical universe was dependent upon higher dimensional realities:

"Oh WRETCHED race of men, to space confined! What honour can ye pay to him, whose mind To that which lies beyond hath penetrated? The symbols he hath formed shall sound his praise, And lead him on through unimagined ways To conquests new, in worlds not yet created.

First, ye Determinants! In ordered row And massive column ranged, before him go, To form a phalanx for his safe protection. Ye powers of the nth roots of - 1! Around his head in ceaseless* cycles run, As unembodied spirits of direction.

And you, ye undevelopable scrolls! Above the host wave your emblazoned rolls, Ruled for the record of his bright inventions. Ye cubic surfaces! By threes and nines   Draw round his camp your seven-and-twenty lines- The seal of Solomon in three dimensions."

The 21st Century evidence is mounting. The universe is not only expanding, but at a rate that accelerates with each passing moment. Distances between two points are measured now in AU's—Astronomical Units are the distance between the Earth and the Sun = 93 million miles—instead of light years. Recent experiments have shown that matter can indeed move faster than the speed of light. In fact, particles have been repeatedly clocked at speeds above 300 times the speed of light. Contrary to accepted mathematical limits—$E=mc^2$—mass did not become infinite.

There is a place of intersection between the libraries and universities full of rigid thinkers and the pure and simple knowledge man seems to have had since the first stylus was laid to clay or gold. There are indeed many dimensions. Mathematicians and physicists are catching up to prophets. As they do, we will all be able to read and scribble out not only the proof, but will be able to insert our minds and perhaps our bodies into the process and really discover the universe. Like fish in the sea, man has innocently imagined his realm beneath the ionosphere as the whole of the universe. Right now we can only look at lights, whose realities are millennia away and older than the earth itself. Right now, if there were a planet belched from the maelstrom of the galaxy by a nova or a rupturing black hole at the speed of light, we would not be granted the time to even pray before its collision with our solar system. It would physically arrive shortly after its image reflected in the mirrors of our most powerful telescopes. The ocean's sentient occupants steal glimpses of the stars through a calm sea and may circumnavigate the globe, but never dream of moving beyond the rich liquid bounty through which they fly. The intersection between the air and the surface of the sea is the boundary of their universe. They are unaware of cities and farms and classrooms. Does this mean they don't exist. No. If a dolphin from Sea World was educated in all the ways of man, including space travel and hypersonic flight through the air, and then returned to the sea where he could pass along this great knowledge to other dolphins would he be labeled a heretic? Would he be followed as a prophet or honored as a great explorer? What would his knowledge of the hyper-universe be worth to his dolphin brothers? Would he forever leap into the air and strain to see a plane or a satellite or the bright lights of the city?

The vast majority of people wander through life asleep; unaware of the hyper-universes above them. But, there are a few of us who are awake, and live in a constant state of amazement. To be able to jump between the threads of *the bed sheet universe* may allow us to see these universes, and thus the explanation of the past of our existence. If small particles can be accelerated to many times the speed of light, or pass between dimensions like a universe made like a hall of mirrors, then it too must be possible for larger particles, such as planets, to do so as well. Every human who correctly arrived at the conclusion that there is a creator God also knows that our ignominious accomplishments do not give us the authority to disclaim this possibility. The water planet described in the *Genesis* creation long ago was divided.[3] It's daughter was moved in its clean and liquid state through the firmament to a place where the sun rules the day and the moon rules the night by passing from one dimension to another.

We know that black holes exist. We have recent and compelling evidence that Einstein's theories about black holes are fairly accurate. Through decades of contemplation and long suffering detectors have been engineered to provide data to test these theories. Piece by piece we have proven some things to be true. And, we have discovered even more than we imagined.

Black holes don't sit still in space. In fact, if one thinks about it, what does it

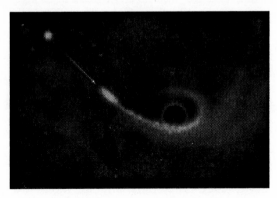

[In this picture one can see a star being ripped apart by a spinning black hole.] mean to be still in space? Relative to what? There is no absolutely motionless spot in space. The whole of the known universe, and most assuredly those we suspect exist beside the one we think we occupy, is one infinite Heisenberg Uncertainty. The absolute position of anything cannot be predicted or detected without altering the very thing we are trying to detect. Black holes move through space. Black holes rotate, sending immense gravitational waves out into space.

We know that mass moving through space warps space-time, and black holes have tremendous mass in a very compact area. Picture a bedsheet pulled tight by a person in each corner. Then, picture a bowling ball being placed in the center of the sheet. The threads of the sheet will warp around the surface of the bowling ball just like the space-time continuum warps around any mass moving through space. The amount of warp in the bedsheet depends on what weight and diameter bowling ball we use.

We also observe that when objects spin, the portion of their mass forced away from the center of rotation is proportional to its mass and the speed of rotation. In other words, the faster and heavier an object is, the more force it will exert against the gravity at the center. The same thing occurs when a black hole spins. It is our contention that scarcely a black hole will be found that does not spin. The faster a black hole spins, angular momentum takes over and its sphere becomes squashed, and the further from the center that mass is thrown, and the larger the proportion of the object's mass is thrown toward the outside boundaries of the spinning plane.

Take that to the logical conclusion and you may observe a spinning black hole 400 times the diameter of our solar system with 90% of its matter on the outer rim and only 10% of its matter spread across the rest of the diameter of that plane. In the center of the plane there may be as little as 1% of the total mass of the black hole strewn over 50 million miles of space. The distortion of space-time from such a black hole would couple countless universes together. The black hole's singularity—the point at which all movement stop— would be divided across the centrifugal body. As long as a traveler avoids reaching a singularity, it is not destroyed. The traveler passes from one event horizon—the threshold at which a traveler enters the black hole—to the next and on through the black hole without damage. An experienced traveler could move forward or backward in time from universe to universe through such a portal. Roy Kerr presented this algebraic possibility in 1985, long before the Hubble telescope displayed the possibilities of Einstein's imagination.

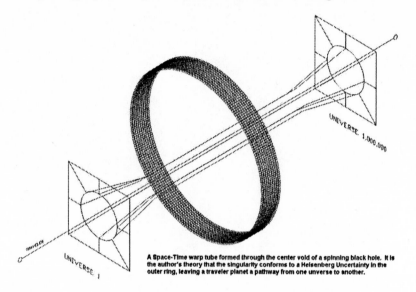

A Space-Time warp tube formed through the center vold of a spinning black hole. It is the author's theory that the singularity conforms to a Heisenberg Uncertainty in the outer ring, leaving a traveler planet a pathway from one unverse to another.

This is the foundation for the theory of transdimensional space travel. Imagine a free-floating planet moving through space on a course that will take it through the center of the black-hole ring (see figure above) perpendicular to the plane of its rotation. That planet may suffer minor loss of mass as the outer ring of the black hole exerts its gravity, far enough from Roche limits—the distance at which the gravitational diameter of the larger overcomes the gravitational diameter of the smaller body, resulting in internal forces shattering the smaller body—to spare the planet from any real structural damage.

A path through the sparsely populated epicenter of the spinning black hole may facilitate travel far across the universe while the traveling planet barely ages. In other words, the planet may warp across huge distances of space without any damage, and seemingly in an instant of time, from the point of view of the traveler. The geomagnetic forces of black holes are being observed by the most modern instruments available, and are a marvel to be sure.

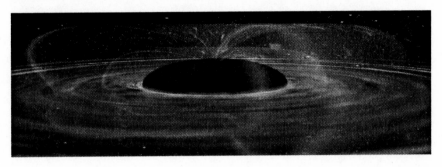

The point of observation to factually verify that a planet may warp across space and time has never been achieved, except through the imagination of fellow theoretical physicists. It would be nearly impossible to position one's self at the edge of the spinning black hole, much like looking across the bedsheet from the edge, while a planet approaches and passes through from one side to the other. Even at a thousand miles per second, it may take months, or even centuries, for the journey to occur, from the point of view of an outside observer. What we want you to think about is the possibility for a planet to be transported across great expanses of time and space with no more difficulty than an electron transports from one uncertain spot to another around an atom. Whether by divine direction or mere cosmic chance, depending on the camp at which you warm your hands, the possibility appears to exist. At least this physicist would love to one day, in future eternities, find out if this does in fact occur.

The completed Earth will one day return to its origin, according to the prophets. It will receive it paradisiacal glory. The spirit of the earth will be separated from the physical earth and return home. The union of the polarity will be complete. The heavens shall roll together as a scroll, and the earth shall pass away. [4] Sound familiar?

The mathematics it takes to describe these quantum changes on a scale larger than two protons is not what this book is about. The concert playing across our telescopes contains countless vibrations. Some we can see with our eyes. Some we can hear with our ears. The ones that stray into our radio, x-ray, infrared, or gravitational telescopes are but a spec of sand in the Sahara compared to the truth of the universe. The authors want you to think about that. The intention is not to make the reader feel small, but to inspire the mind to stretch and comprehend that your potential is absolutely infinite, limited only by your ability to receive truth.

There are beings that know how the universe was made. They may indeed be in the process of making it still. If you believe you are a descendant of those beings then your mind is authorized to think well beyond the surface of the planet from which we pray. Think beyond the rocket-reached system imagined upon each cloudless night for guidance and inspiration. Open your mind to imagine that billions of solar systems lay beyond even your most powerful gaze. Believe that you were meant to know about them.

Perhaps civilizations, long since vanished from this planet, may have known far more than our most brilliant scientists could hope. When we find them, will we destroy their records in fear? Will we bury their cities and melt down their books like forbidden fairy tales? Will we ignore them unless canonized by some robed authority? Will we call them a cult and close our minds to the possibility that the Creator spoke to someone other than our earthly affiliate? What fish-like absurdity.

The writers are begging you to leap into the sky one more time. Grab hold of our hands. In the next few chapters we will carry you to a place where the scientist and the prophet dwell in harmony. We can hardly wait!

End Notes

[1] *How Gravitational Waves Lead to Space-Time Ripples*
  By <u>*Andrew Bridges*</u> *Chief Pasadena Correspondent*
  posted: 12:46 pm ET
  15 November 1999
  Chief Pasadena Correspondent
  posted: 12:46 pm ET
  15 November 1999
[2] Et al: *The Acts*, chapter 8, verses 39-40 & chapter 10, verses 11, 16.
[3] et al *Isaiah* 4:4;
[4] *Revelation* 6:14

The Heavens

According to ancient writings, beliefs, and sacred scriptures, the heavens are many in number. In *The Holy Bible* it says:

"I knew a man in Christ above fourteen years ago, (whether in the body, I cannot tell; or whether out of the body, I cannot tell; God knoweth;) such as one caught up to the third heaven"[1] Where there is a third heaven, there is a second heaven, and where there is a second heaven, there is a first heaven.

King Solomon states during the dedication of the temple:

"But will God indeed dwell on the earth? behold, the heaven and heaven of heavens cannot contain thee; how much less this house that I have builded?"[2]

"In the beginning God created the heaven and the earth" *(Genesis 1:1)*

Notice that in the last verse, the King James version *Holy Bible*, in English, the word heaven is singular. Yet, the word *heaven* is translated from the Hebrew word shamaim=shawmah-eem, which is the plural form. In fact, in every reference in the Hebrew texts, there are no singular references for heaven. It is always plural.

In the *Doctrine and Covenants*, more insight is given on these three heavens. [3]

89 "And thus we saw, in the heavenly vision, the glory of the telestial, which surpasses all understanding;"

91 "And thus we saw the glory of the terrestrial which excels in all things the glory of the telestial, even in glory, and in power, and in might, and in dominion."

92 "And thus we saw the glory of the celestial, which excels in all things----where God, even the Father, reigns upon his throne forever and ever."

In the latter record we learn the names of the three heavens; the telestial, the terrestrial, and the celestial. The telestial is the lowest, often referred to as hell in many writings. The terrestrial is a heaven reserved for those who lived good lives, but could not accept a fullness of the gospel. The celestial heaven is the third heaven and the one in which God, often called heavenly Father, resides. The celestial heaven contains three levels of glory as well. [4] These levels of glory could be interpreted to mean that the celestial glory is divided into three other heavens or universes.

Moses inquired of the Lord, "Tell me concerning…. the heavens" *(Moses 1:36,37)* The Lord answered, and as Moses gazed upon these heavens and worlds the Lord explained that "the heavens, they are many, and they cannot be numbered unto man; but they are numbered unto me, for they are mine."

In the *Apocrypha, The Wisdom of Solomon,* describes multiple heavens.

"O send her (wisdom) out of thy holy Heavens and from the throne of thy glory that being present she may labour with me, that I may know what is pleasing unto thee."[5]

In the *Pseudepigrapha, The Ascension of Isaiah,* gives an extensive account of *Isaiah* ascending to seven heavens and what he saw in each heaven. One mention of multiple heavens, taken from that account:

13 "And the angel who was sent to show him (the vision) was not of this firmament, nor was he from the angels of glory of this world, but he came from the Seventh Heaven."[6]

In *Old Testament Pseudepigrapha, The Revelation of Moses,* the following two accounts are given:

569 "Look up with thine eyes, and see the seven firmaments (heavens) opened, and see with thine eyes how the body of thy father (Adam) lies upon its face, and all the holy angels with him, praying for him, and saying; Pardon him, O Father of the universe; for he is Thine image."

"And then He said to the archangel Michael: Go into paradise, into the third heaven, and bring me three cloths of fine linen and silk."[7]

The *Pseudepigrapha* gives another account of multiple heavens in the *Secrets of Enoch*. Here Enoch, taken by angels, visits ten heavens, and he reveals what is seen in each of those heavens. [8]

The *Pseudepigrapha, Testament of Levi*, tells of Levi, third son of Jacob and Leah, journeying to three different heavens with an angel guide and relating his experiences and what he saw in each heaven. [9]

The *Pseudepigrapha* writings of *The Ascension of Isaiah* state: "And the angel who was sent to show him (the vision) was not of this firmament, nor was he from the angels of glory of this world, but he came from the seventh heaven."[10]

In the ancient Heilhalot literature, *The Riders of The Chariot*, we read:

"Thus Ezekiel stood beside the river Chebar gazing into the water and the seven heavens were opened to him so that he saw the Glory of the Holy One, blessed be He, the hayyot, the ministering angels, the angelic hosts, the seraphim, those of sparkling wings, all attached to the merkavah."

They passed by in heaven while Ezekiel saw them (reflected) in the water. Hence the verse says: "by the river Chebar, that the heavens were opened" teaches that seven heavens were opened to Ezekiel: Shamayyim; Shemei ha-Shamayyim; Zevul; Araphel; Shehakim; Aravot; and the Throne of Glory. Rabbi Meir said: The Holy One, blessed be He, created seven heavens and in these there are seven chariots (merkavah). [11]

Jewish Kabbalah teaches that there is a hierarchy of worlds and the teachings are the central heart and core of Jewish mysticism.

According to Lurianic Kabbalah, there are five different worlds, beginning with the Godhead and ending with the lowest, but also intrinsically spiritual world. These are: Adam Kadmon (i.e., the Sefirotic world), the World of Atsiluth or Emanation; the World of Beri'ah or Creation (also the World of the Divine Throne); the World of Yetsirah of Formation (i.e., the angelic world); and the World of Assiyah (perhaps best translated as "work completion" or "concretization").[12]

The earlier Jewish Kabbalah of the *Sefer Yetzirah* (*The Book of Creation*) teaches that there are four universes or heavens. The highest universe is Atzilut or Emanation, the world closest to the Throne of Glory and the domain of the Sefirot themselves. Below this is Beriyah, the World of Creation. Below Beriyah is the universe of Yetzirah (formation), which is the world of the angels. Finally, there is the universe of Assiyah (Making), which consists of the physical world and its spiritual shadow. It is the universe in which we live. [13]

According to the *Sefer Yetzirah,* the heavens consist of five dimensions. The space continuum consists of three dimensions, up-down, north-south, and east-west that is called "Universe." The time continuum consists of two directions, past and future, or beginning and end. This is called "year" or the fourth dimension of space-time. Finally, there is a moral, spiritual fifth dimension, whose two directions are good and evil. This dimension is called "soul."[14]

In theory if we could access the fourth dimension then space travel would be feasible as well as knowing the past, present and future. Modern physicists calculate it takes ten dimensions to make Einstein's equations become stable. Only now are computers powerful enough to work the models.

In the – *Qur'an* the following writings tell of seven heavens that were created by Allah.

2. The Cow (Al-Baqarah)

002. 029 "He it is Who created for you all that is in the earth. Then turned He to the heaven, and fashioned it as seven heavens. And He is knower of all things."

17. The Night Journey (Al-Isra)

017. 044 "The seven heavens and the earth and all that is therein praise Him, and there is not a thing but hymneth His praise; but ye understand not their praise. Lo! He is ever Clement, Forgiving."

23. The Believers (Al-Muminun)

023. 086 Say: "Who is Lord of the seven heavens, and Lord of the Tremendous Throne?"

41. Fussilat

041. 012 "Then He ordained them seven heavens in two Days and inspired in each heaven its mandate; and We decked the nether heaven with lamps, and rendered it inviolable. That is the measuring of the Mighty, the Knower."

65. Divorce (At-Talaq)

065. 012 "Allah it is who hath created seven heavens, and of the earth the like thereof. The commandment cometh down among

them slowly, that ye may know that Allah is Able to do all things, and that Allah surroundeth all things in knowledge."

67. The Dominion (Al-Mulk)

067. 003 "Who hath created seven heavens in harmony. Thou (Muhammad) canst see no fault in the Beneficent One's creation; then look again: Canst thou see any rifts?"

71. Noah (Nuh)

071. 015 "See ye not how Allah created seven heavens in harmony."

The ancient Maya Indians or Nahua also believed in three different heavens which their dead were suppose to go. These heavens also have many records throughout the *Popol Vuh*. Modern researchers have also found proof of a mythological belief that corroborates these records found throughout Mexico and the Yucatan.

The first and lowest of these was Tlalocan, Land of Water and Mist; a kind of paradise where happiness was of a very earthly variety but purer and less changeable. Here people played leapfrog, chased butterflies and sang songs. One fresco from Teotihuacán, ancient Mayan ruins, later occupied by the Aztecs, outside of Mexico City, depicts this scene in a charmingly light-hearted manner.

There was also Tlillan-Tlapallan, the land of the black and red (black and red in conjunction signify wisdom). This was the paradise of the initiates who had found a practical application for the teaching of the god—king Quetzalcoatl. It was the land of the fleshless, the place where people went who had learned to live outside their physical bodies or, it would be better to say, unattached to them; a place celebrated in many ancient poems and greatly to be desired. The highest heaven was referred to as the thirteenth heaven, or the paradise of Tamoanchan.

Further beyond was Tonatiuhican, House of the Sun. The third paradise was probably reserved for those who had achieved full illumination in the quest for deserved eternal happiness.[15]

In Mahayana Buddhism there are five Cosmic Buddhas and each presides over a heaven. The Cosmic Buddha, Amitabha, is Lord of the Western Paradise. Buddhist descriptions of Amitabha's heavenly realm involve three degrees of glory. The first degree is described as being reserved for "those who have lived steadfast in purity." The second degree is for "others who have led devout lives as laymen." And the lowest degree is for "those whose good impulses have been offset by backsliding or unbelief." It is only in the topmost of these that the faithful will fully enjoy the presence of deity. And finally, "all these three degrees (of glory) will share equally in the bliss of the Western Land, in the end; only the last group will have to pass through a purgatorial period of five hundred years.... The souls on probation will be isolated, unable to see the Buddha or hear Him preach until their term has run out."[16]

In the *Gnostic scriptures, Eugnostos* (111,3), many more heavens are described.

"And in each aeon there were six heavens so there are seventy-two heavens of the seventy-two powers who appeared from him. And in each of the heavens there were five firmaments so there are altogether three hundred sixty (85) (firmaments) of the three hundred sixty powers that appeared from them."

"When the firmaments were complete, they were called "The Three Hundred Sixty Heavens," according to the name of the heavens that were before them. And all these are perfect and good."[17]

The *Gnostic scriptures of The Apocryphon of John* (11,1, 111,1, 1V,1, and BG 85502,2) state:

"And he placed seven kings—each corresponding to the firmaments of heaven—over the seven heavens, and five over the depth of the abyss, that they may reign."[18]

The *Gnostic scriptures of Zostrianos* (V111,1), opens with an account of Zostrianos journeying through multiple heavens in pursuit of saving gnosis or knowledge. The physical earth and its inhabitants represent the lowest and most ignorant level. Above the lowest level lies a vast aeon (space) system called Barbelo that is divided into three constituent aeons: the upper most aeon is named Kalyptos (the hidden or veiled aeon), in the middle is located the Protophanes (the first-visible or first-appearing aeon), and at the bottom is the Autogenes (the self-generated or self-begotten aeon). Each of these aeons, or heavens, has its own system of constituent beings called lights, glories, angels, waters, etc.[19]

The Gnostic scriptures of *The Apocalypse of Paul* (V,2), gives another account of Paul being caught up on high to the third heaven and continues to ascend to the fourth, fifth, sixth, seventh, eighth, ninth and tenth heaven. He gives an account of what he sees in each heaven.[20]

From *the Messianic and Visionary Recitals*, in *The Words of Michael* (4Q529), text of the *Dead Sea Scrolls*, a mention of multiple heavens appears. Here the Archangel Michael ascends to the Highest Heaven and then descends to tell the ordinary Angels what he has seen. Some practitioners of this kind of mystic journeying speak of "three layers", some of seven, and some of twelve (heavens).[21]

An ancient Chinese creation myth titled *Nuwa Mends the Sky* states as follows:

"She took ghosts and Gods to the ninth heaven and had an audience with the Heavenly Emperor at Lin Men."

The Zohar states, "There are lower heavens and upper heavens, both referred to in the passage, saying, "The heavens, heavens of the Lord,"[22] but this is the supremest heaven, raised over them all."[23]

Adin Steinalz writes in his opening lines of *The Thirteen Petalled Rose:* " The physical world in which we live, the objectively observed universe around us, is only a part of an inconceivably vast system of worlds. Most of these worlds are spiritual in their essence; they are of a different order from our known world."

The word heavens is usually thought of as planets, moons, and stars, seen from the earth, but there are other unseen dimensions that contain multiple heavens and universes full of worlds. These dimensions are also called heavens and are the heavens that the above scriptures, ancient writings, beliefs, seers and prophets all refer to when they speak of the "heavens". These dimensions, or heavens, contain universes that also contain worlds, without end, and are the heaven of heavens that King Solomon referred to during the dedication of the temple.

In the *New Testament* of *The Holy Bible*, John 14:2, it states, "In my Father's house are many mansions." In other words—"In my Father's house there are many worlds." Heaven is not one world but many.

If the reality of multiple heavens is irrefutable, each a different level of glory than the next, then the quality of members of those heavens must differ as well. The various scriptures seem to indicate that conformity with the rules or commandments of the ruler of that heaven determine who lives there and who does not. Good members live here. Bad members live there. There seems to be evidence, through witness or participation, of humans receiving the privilege of seeing from one heaven to another.

There are many other records that indicate an almost universal belief in multiple heavens. Hardly a civilization can be researched whose prophets or historians wrote of a common belief in many degrees of glory in the universes. Note the plural of universe. Each of these heavens is a universe in itself, to be addressed in the next chapter. Just how many heavens in number exist are not known at present but certainly some have been revealed to us to indicate that the heavens indeed are many, and certainly more than one.

End Notes

[1] *(2 Corinthians 12:2)*

[2] (*1 Kings* 8:27)

[3] (*D&C* 76:89, 91,92)

[4] (et al *D&C* 131:1)

[5] (*The Missing Books of the Bible*, Vol. 1, p. 214 Ottenheimer Publishers, Inc. 1996)

[6] (online text www. Pseudepigrapha. com)

[7] (online text www. Pseudepigrapha. com)

[8] Excerpted from *The Forgotten Books of Eden* ©1927 by Alpha House, Inc. Published by World Publishing, Nashville, TN. Used by permission of the publisher. All rights reserved. pp.82-89 copyright in the United States of America and in Newfoundland 1927 by Alpha House, Inc. Published by World Bible Publishers, Inc.)

[9] (Excerpted from *The Forgotten Books of Eden* ©1927 by Alpha House, Inc. Published by World Publishing, Nashville, TN. Used by permission of the publisher. All rights reserved. , the *Testament of Levi*, p. 227, copyright in the United States of American and in Newfoundland 1927 by Alpha House, Inc. , Published by World *Bible* Publishers, Inc.)

[10] www. pseudegrapha. com

[11] From **JEWISH MYSTICAL TESTIMONIES** by Louis Jacob, copyright ©1977 by Keter Publishing House Jerusalem, Ltd. Used by permission of Schocken Books, a division of Random House, Inc page 38-39

[12] From **ON THE MYSTICAL SHAPE OF THE GODHEAD** by Gershom Scholem, English translation copyright ©1991 by Schocken Books. Used by permission of Schocken Books, a division of Random House, Inc. page 230

[13] (Excerpted from *Sefer Yetzirah* by Aryeh Kaplan ©1997 with permission of Red Wheel/Weiser, York Beach, ME and Boston, MA p. 43)

[14] (*Sefer Yetzirah* , et al p. 50)

[15] (see *Mexican and Central American Mythology*, by Irene Nicholson, pp.25-26)

[16] (Religions of The World, p. 59)

[17] (*The Nag Hammadi Library*, James M. Robinson general editor Harper Collins 1990; Eugnostos (111. 3) italics added. P. 236)

[18] (*The Nag Hammadi Library*, James M. Robinson general editor Harper Collins 1990; The Apocryphon of John (11,1 111,1, 1V,1, and BG85502,2) p. 111

[19] (The Nag Hammadi Library, James M. Robinson general editor Harper Collins 1990; Zostrianos (V111,1) p. 402)

[20] (*The Nag Hammadi Library*, James M. Robinson general editor Harper Collins 1990; The Apocalypse of Paul (V,2) pp.258-259)

[21] *The Dead Sea Scrolls Uncovered*, by Robert Eiseman & Michael Wise, pp.37-38)

[22] ((Ps. CXV, 16)

[23] (*Zohar* Vol. 1V, translated by Maurice Simon, Dr. Paul P. Levertoff, The Soncion Press, N.Y., 1984, p. 224)

NOTE: The King James Version [KJV] of the *Holy Bible* is used throughout this book, unless otherwise noted

The Eternal Heavens

The word eternal means to exist forever in an endless, everlasting, imperishable or unchanging steady state. The universe that we live in is not eternal but temporal in existence. This means that our universe will not, does not, and cannot continue to exist forever. It had a beginning and it will have an ending. It does not obey an unchanging steady state law as once thought. It boggles our mind to know that entire galaxies can and have exploded. Luminous nebulae are enormous clouds of dust and gas that are busy breeding grounds for planet and star formation. Thanks to the Hubble telescope, those mysterious objects called quasars are now thought to be instrumental in galaxy formation. Aging stars eventually turn into supernovas, explode and die. Black holes can gobble up matter and hundreds of stars per year. Anything they trap can never escape, even light. We can now better understand the Lord's statement to Moses, when he said:

"And as one earth shall pass away, and the heavens thereof, even so shall another come."[1]

The eternal heavens that the sages, prophets, seers and revelators all speak of are not located in our universe because this universe is temporal or subject to laws of change. The eternal heavens remain unchanged forever and cannot reside in temporal abodes; to do so would make them subject to change, and would eventually perish.

If these heavens do not exist in our universe, where are they located? They are located above our universe in another dimension, called the fifth dimension or "soul." We can safely say that there are at least four heavens, if you count our physical universe as the lowest heaven, and in all probability more, ranging from five to ten, and possibly even more. Most all sacred writings generally accept the concept of seven heavens. These heavens are placed one above another and are vast regions called aeons. The aeons are immense universes containing galaxies, stars and planets. They are called the spiritual creations or spiritual heavens. Our physical universe was patterned after a spiritual universe.

The *Qur'an* states in Nuh:

(71. 15) "Do you not see how Allah has created the seven heavens, one above another."

The Kingdom, also from the *Qur'an* states:
(67. 3) "Who created the seven heavens one above another; you

see no incongruity in the creation of the Beneficent God; then look again, can you see any disorder?"

In *The Book of Alma*, this scripture reads: And Ammon said unto him: "The heavens is a place where God dwells and all his holy angels.

And King Lamoni said: Is it above the earth?

And Ammon said: Yea...."[2]

The *Pseudepigrapha, Secrets of Enoch*, states:

"The Lord has placed the foundations in the unknown, and has spread forth heavens visible and invisible....

From the invisible he made all things visible, himself being invisible."[3]

The Gnostic scriptures of *The Hypostasis of The Archons* state:

"And she (Faith/Wisdom) established.... in conformity with its power—after the pattern of the realms
that are above, for by starting from the invisible
world the visible world was invented."[4]

The Gnostic scriptures of *Zostrianos* relate how Zostrianos never tires of asking his angelic revealers about how this changeable world came into existence from an unchanging source, or about why there are different kinds of souls or animals or human beings. Their answers assert that each successive layer of the universe is formed on the basis of the models in the layer above and that each layer is less perfect than its model.[5]

The Gnostic scriptures of *Zostrianos* explain how the creations were patterned after one another by stating that; "With a reflection of a reflection he worked at producing the world." (see chapter on *The Dual Creations*). A reflection of a reflection can best be demonstrated by placing mirrors in front of each other. They each reflect each other in endless succession, like looking into eternity.

Another statement from the teachings of Zostrianos state that the aeons (heavens) were copies of each other and that there was a physical or temporal creation as well; "I came down to the aeon copies....... then I came down to the perceptible world and put on my image."[6]

The Gnostic scriptures of *The Teachings of Silvanus* (V11,4) state: "For everything which is visible is a copy of that which is hidden."[7]

A reference that these aeons contain eternal suns is found in *Testaments and Admonitions*, text of the *Dead Sea Scrolls*, which states: "His eternal sun shall burn brilliantly."[8]

The Riders of The Chariot—ancient Heikhalot literature— tells of the seven heavens and the seven thicknesses between them. It further states that the distance from the earth to the firmament (first heaven) is a journey of five hundred years, and likewise the distance between one firmament and the other.[9]

The Diamond Sutra, from Buddhist text, refers to more heavens than seven which says: (13B) "What think you, Subhuti? Are the atoms of dust that comprise the three thousand great universes very numerous?"

"Very numerous indeed, Blessed Lord!"

(8) "What think you, Subhuti? If a disciple bestowed in charity an abundance of the seven treasures sufficient to fill the three thousand great universes, would there accrue to that person a considerable blessing and merit?"[10]

The afore mentioned writings and sacred scriptures all describe the spiritual heavens located above our universe. **Are there spiritual heavens located below our universe?** The following sacred scriptures and writings indicate there is at least one aeon (heaven) below our universe, and perhaps more. This could be the placement of the First Heaven, or lowest degree of glory, referred to by some as Hell or another spiritual realm of which little is known.

The Holy Bible refers to such a place in the following scriptures:

"Hell **from beneath** is moved for thee to meet thee at thy coming: it stirreth up the dead for thee, even all the chief ones of the earth; it hath raised up from their thrones all the kings of the nations."[11]

"The way of life is above to the wise, that he may depart from **hell beneath**."[12]

"They also went **down into hell** with him unto them that be slain with the sword;......"[13]

"And no man in heaven, nor in earth, neither **under the earth**, was able to open the book, neither to look, thereon."[14]

"And every creature which is in heaven, and on the earth, and **under the earth**, and such as are in the sea, and all that are in them, heard I saying, Blessing, and honour, and glory, and power, be unto him that sitteth upon the throne, and unto the Lamb for ever and ever."[15] This scripture implies that there are creatures or animals in those spiritual realms.

"That at the name of Jesus every knee should bow, of things in heaven, and things in earth, and things **under the earth**;"[16]

In *The Book of Moses* we read:

"And behold, all things have their likeness, and all things are created and made to bear record of me, both things which are temporal, and things which are spiritual; things which are in the heavens above, and things which are on the earth, and things which are in the earth, and things which are **under the earth**, both above and **beneath**: all things bear record of me."[17]

From the *Doctrine and Covenants*:

"And this shall be the sound of his trump, saying to all people, both in heaven and in earth, and that are **under the earth-** --for every ear shall hear it, and every knee shall bow, and every tongue shall confess, while they hear the sound of the trump, saying: Fear God, and give glory to him who sitteth upon the throne, forever and forever; for the hour of his judgment is come."[18]

From the *Qur'an* in the Fussilat 041.012:

"Then He ordained them seven heavens in two Days and inspired in each heaven its mandate; and We decked the **nether heaven** with lamps, and rendered it inviolable."

(bold type emphasis added to each scripture)

One of the most ancient texts, dating back to at least 3,500 B.C.E., is the Sumerian Text which recants numerous visits to a region beneath the earth called the Nether World. Kramer, in his book, *The Sumerians*, states, "This region was divided into seven realms and ruled by deity. The Sumerian thinkers, in line with their world view, believed that when man died, his emasculated spirit descended to the dark, dreary nether world where life was but a dismal and wretched reflection of its earthly counterpart. In general the nether world was believed to be the huge cosmic space below the earth corresponding roughly to heaven, the huge cosmic space above the earth. Although one has the feeling that the nether world was always dark and dreary, this would seem to be true only of daytime; at night the sun brought light to it, and on the twenty-eight day of the month the sun was joined by the moon."

The Sumerian document which provides the most detailed information about the nether world and the life going on within its realms, is the poem "*Gilgamesh, Enkidu, and the Nether World.*"[19]

The Papyrus of Ani, better known as *The Egyptian Book of The Dead*, is the best preserved, and the best illuminated of all the papyri which date from the second half of the XV111th dynasty (about 1500 to 1400 B.C.E.) Wallis Budge in his work, *The Egyptian Book of the Dead*, states the following, "The Egyptians believed in an underworld called Taut wherein dwelt the gods of the dead and the departed souls."[20]

The Taut

This view is supported by the scene from the sarcophagus of Seti 1. In the watery space above the bark is the figure of the god bent round in a circle with his toes touching his head, and upon his head stands the goddess Nut with outstretched hands receiving the disk of the sun. The two bowed female figures represent the day and the night sky and the third figure which is bent round in a circle, the space enclosed in the Taut. In a similar space enclosed by the body of the god is the legend, "This is Osiris; his circuit is the Taut." Budge continues, "Osiris as a legendary character as the god of the dead was well defined long before the version of the pyramid texts known to us were written. The ceremonies connected with the celebration of the events of the sufferings, the death and the eventual resurrection of Osiris played a prominent part in the religious observances of the Egyptians in the month of Choiak that took place in various temples in Egypt; the text of a minute description of them has been published by M. Loret in Recueil de Travaux, tom. 111, p. 43 ff., and succeeding volumes."

"A perusal of this work explains the signification of many of the ceremonies connected with the burial of the dead, the use of amulets, and certain parts of the funeral ritual; and the work in this form being of a late date proves that the doctrine of immortality, gained through the god who was "lord of the heavens and of the earth, of the underworld and of the waters, of the mountains, and of all which the sun goeth round in his course," had remained unchanged for at least four thousand years of its existence." One text, from the book, reads: An-maut-f saith: "I have come unto you, O mighty and godlike rulers "who are in heaven and in earth and under the earth; and I have brought unto you Osiris Ani."[21]

The *Vision of Ezekiel* text from *The Riders of The Chariot* tells how the Holy One, blessed be He, opened seven compartments down below. Ezekiel gazed into these in order to see all that is on high. These are the seven regions below: Adamah ("ground"); Erez ("Earth"); Heled ("World"); Neshiyyah ("Forgetfulness"); Dumah ("Silence"); She'ol ("Pit"); and Tit ha-Yaven ("Miry Clay"). It is interesting to note that Ezekiel saw his visions of the higher worlds or heavens by gazing down into the divisions of the netherworld.[22]

How are these heavens, universes, or aeons constructed? As mentioned before, these heavens are layered one on top of another. How far are they stretched out is not presently known, but the scriptures and ancient writings attest that the organized universes are countable to the creator and are stretched out. This was explained to Moses, by the Lord, in Moses' vision of the universes:

"There are many worlds which have passed away....and there are many that now stand, and innumerable are they unto man; but **all things are numbered unto me**, for they are mine and I know them."[23]

"The heavens, they are many, and they cannot be numbered unto man; **but they are numbered unto me**, for they are mine."[24]

"The Prophet *Isaiah* refers to the expansion of the boundaries of the organized universes when he speaks of "the Lord, thy maker, that hath **stretched** forth the heavens."[25] (bold type added)

The ancient Heilhalot literature in *The Riders of the Chariot*, states that, "these heavens are fashioned in no other way than as a dome."[26] Another way of describing it is almost spherical in shape. The teachings of Islam also supports this in Allah's saying:

"His Kursi extends over and encompasses the heavens and the earth", supports the opinion held by Shaykh-ul-Islaam ibn Taymeeyah and by other verifying scholars that the heavens and the earths all have a spherical shape. The fact that al-kursi extends over and encompasses the heavens and the earth is evidence that it is wound round in a round form. Regarding al-'arsh, it has been reported that the Prophet described it as being like a dome above the heavens. The dome-shape being round but neither fully spherical nor flat, and its middle is high like that of a tent.

From *Isaiah*: "It is he that sittest upon the circle of the earth, and the inhabitants thereof are as grasshoppers; that stretcheth out the heavens as a curtain, and spreadeth them out as a tent to dwell in."[27]

The Teli is the axis, or hub, around which all the universes revolve. In circular motion around an axis, everything moves but the axis itself. The axis is the focus of the motion, but does not partake in it. Similarly, the **Teli** is viewed as the king over the universes, that is, over the domain of space, but does not become part of it. It is the imaginary axis around which all the heavens (universes or aeons) rotate. The word **Teli** comes from the root Talah, meaning "to hang". The great universes, both spiritual and physical, are hung or suspended around the **Teli** and rotate in a heavenly circle. Thus the **Teli** makes firm or calms the heavenly circle. These great universes are immense and so stretched out that the limits have not yet been discovered nor the gradual curve of the heavenly circle around the **Teli.**[28]

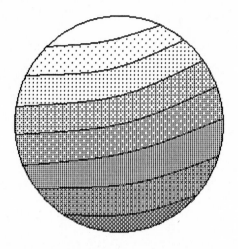

Single stacked balloon universe as imagined by E.J. Clark. Notice the gradient of light or glory from level to level.

According to Einstein, the gravitational pull created by massive bodies can be expressed as a curvature of space. He showed mathematically that the entire universe must have an all-pervading curvature. The more material present, the greater would be the curvature. And the degree of curvature is a measure of whether the universe is slowing its outward rush or growing without limit.[29]

The word stretch means to "extend or expand." Stretch and expand are synonymous. To say the Universe is expanding is to say that it is stretching. Our Universe is expanding at a much faster rate than previously thought but at the same time there is plenty of space in which to expand. There are those who believe that the Universe expands to its limit and then contracts or returns again to near nothingness. This is the so-called oscillation theory.[30] This theory doesn't apply to the spiritual heavens, who may be expanding as well, but remain everlasting. Will our universe eventually self destruct? It is possible since it is a temporal creation. Remember the word temporal means temporary; that it had a beginning and will some time end. The oscillation theory is supported by the Vedic, Buddhism, Janinism, along with Brahmanism views which hold that our present universe is only one in a beginningless series, reappearing and disappearing from time to time. The Buddha is reported to have recovered the memories of eighty-four past world-periods, an immense stretch of time almost impossible of comprehension. [31] Readers are also referred to the chapter, *The Future World.*

The Holy Bible describes the eventual end of this universe in *II Peter* 3:10, and 12 as follows:

"But the day of the Lord will come as a thief in the night; in the which the heavens shall pass away with a great noise, and the elements shall melt with fervent heat, the earth also and the works that are therein shall be burned up."

"Looking for and hasting unto the coming of the day of God, wherein the heavens being on fire shall be dissolved and the elements shall melt with fervent heat?"

How are these great universes separated? In the Gnostic writings of The *Hypostasis of The Archons* it states: " A veil exists between the world above and the realms that are below."[32]

From the *D&C* 88:95, "And there shall be silence in heaven for the space of half an hour; and immediately after shall the curtain of heaven be unfolded, as a scroll is unfolded after it is rolled up, and the face of the Lord shall be unveiled."

When Enoch was permitted to see a vision of the organized universe, he exclaimed: "And were it possible that man could number the particles of the earth, yea, millions of earths like this, it would not be a beginning to the number of thy creations; **and thy curtains are stretched out still!"**[33]

In the fourth volume of *The Zohar—The Book of Splendor—* speaking on the creation, it reads: "At first the light was at the right and the darkness at the left. What, then, did the Holy One do? He merged the one into the other and from there formed the heavens. He brought them together and harmonized them, and when they were united as one, He stretched them out like curtains."[34]

Each great universe is partitioned off by a curtain like structure called a veil. The spiritual universes or aeons are invisible to the natural eye of man. The *Pseudepigrapha* writings of the *Secrets of Enoch* states: "The Lord has placed the foundations in the unknown, and has spread forth heavens visible and invisible:"[35] Adam relates to Eve in the *Pseudepigrapha* of the change that has come over their eyes, the result of the fall: "Look at thine eyes, and at mine, which afore beheld angels in heaven, praising; and they, too, without ceasing. But now we do not see as we did; our eyes have become of flesh; they cannot see in like manner as they saw before."[36]

The invisibility of spirit matter beyond detection of human eyes is explained further in the *D&C* 131:7-8:

"There is no such thing as immaterial matter. All spirit is matter, but it is more fine or pure, and can only be discerned by purer eyes; we cannot see it, but when our bodies are purified we shall see that it is all matter."

Do these great universes (aeons) interact with each other? They do, but it is not fully understood at this time. The veils act as a filter, channeling energy from universe to universe, that keep these realms in sync. This topic will be further discussed in the chapter, *The Interaction of Universes*.

What do these universes contain? As previously mentioned, our universe is patterned after the spiritual universes. The spiritual universes contain eternal planets (earths), moons, stars, suns, and galaxies. Many of the worlds are peopled by spiritual beings and spiritual animals. *Psalms 33:6* says: "By the word of the Lord were the heavens made; and all the host of them by the breath of his mouth."

Moses also beheld these creations in the other universes, as well as our universe, as written: "And he (Moses) beheld many lands; and each land was called earth, and there were inhabitants on the face thereof."[37]

Is it not unreasonable to conclude that there are spiritual plants, trees, and vegetation of all types; including seas, lakes, and rivers of spiritual waters that are habituated inhabited by spiritual fishes, fowl and varieties of spiritual animals to enhance these creations? Remember our planet was patterned after one of the eternal earths.

One mention of spiritual animals in found in *Revelation* 19:14 as follows:

"And the armies which were in heaven followed him upon white horses, clothed in fine linen, white and clean."

From the *Qur'an*, The Children of Israel:

44.44"The seven heavens declare His glory and the earth (too), and those who are in them."

The teachings of Islam indicates the vastness of these immense creations and the greatness of al-kursi is clearly stated in the hadeth in which the Prophet Mohammed said: "The seven heavens and the seven earths by the side of al-kursi are naught but as a ring thrown down in a desert land, and such is al-kursi with respect to al-'arsh (the Throne)".

In addition to all of the above things contained in the eternal heavens, there are vast oceans or seas that are spoken of in almost every sacred writing and scripture. These seas are contained in each of the heavens and are larger than all of the combined oceans of earth. The fact that they are not frozen water indicates that the temperature in each of these universes is warm, moderate or temperate. In the *Gnostic writings of Zostrianos* (V111, 1), the names of some of these spiritual oceans, also called the Living Waters, are identified as: the Water of Life, the Water of Blessedness, and the Water of Existence. God supplies the fountain to these waters.[38]

In the *Secrets of Enoch*, writings of the *Pseudepigrapha*, Enoch describes what he saw in the first heaven as follows: "I looked higher, and saw the ether, and they placed me on the first heaven and showed me a very great Sea, greater than the earthly sea."[39]

In the *Sefer Yetzirah* or *Book of Creation* the names of seven earths (an earth in each heaven) are listed as Adamah, Tevel, Nashiyah, Tzaya, Chalad, Eretz, and Chalad. Besides these earths there are seven seas and seven universes. Your authors note the seven universes are discussed further in the book by later Kabbalists who lacked correct knowledge and understanding of the ancient doctrine and as the authors believe applied incorrect theories about the meaning of the seven universes.[40]

The Zohar Vol. IV states, 'There are heavens and heavens; to wit, lower heavens with an earth beneath them, and upper heavens also having an earth beneath them."[41]

All these great spiritual universes and our own physical universe are organized space. **What is contained in the unorganized universe?** For one, there is an endless amount of unorganized space, not limited by bounds, in which to expand or build future worlds. It is a repository of primal building blocks and chaotic elements, just waiting to be organized, in unlimited quantities. The *D&C* reveals that the elements are eternal: "The elements are eternal, and spirit and element, inseparably connected, receive a fullness of joy." Because the elements are eternal, they cannot be destroyed. They will exist forever.

Brigham Young taught that the unorganized universe lies "beyond the bounds of time and space."[42] The *Sefer Yetzirah* states,

"Even in such a timeless domain, however, there is still a kind of hypertime, where events can occur in a logical sequence. The Midrash calls such hypertime, the "order of time" (*seder zemanim*). The expression of "eternity" denotes the realm outside the time continuum, where the concept of times does not exist at all and the expression "eternity of eternities" (*Adey Ad)* denotes a domain that is beyond even such hypertime."[43]   .

Each universe or heaven, above our universe, is a higher level. Each level contains spiritual beings of higher orders or more spiritually perfected than levels beneath them. Paradise is said to be located in the 3rd heaven, universe or aeon. Many scriptures attest that God resides in the highest heaven.

We live in the latter days, a time where knowledge abounds, is more accessible, and better understood. It is a marvelous age of science and technology, space exploration and medical wonders. It is a time spoken of, by the prophets of old, where knowledge, perception and understanding of everything will increase, including knowledge of the entire creation.

The eternal heavens are veiled from us and considered by many to be one of the great mysteries of God that should be left alone; however there is enough information in the sacred scriptures and ancient writings, when compiled together, to give us additional insight of what is contained in them, and how they are structured. If God didn't want us to know, then he would have sealed those scriptures and writings as well.

End Notes

[1] *(Moses 1:38)*

[2] (*Book of Mormon,* Alma 18:30-32)

[3] Excerpted from *The Forgotten Books of Eden* ©1927 by Alpha House, Inc. Published by World Publishing, Nashville, TN. Used by permission of the publisher. All rights reserved. p. 98.

[4] (*The Nag Hammadi Library,* James M. Robinson general editor Harper Collins 1990; *The Hypostasis of The Archons,* p. 163)

[5] (*Nag Hammadi Library,* James M. Robinson general editor Harper Collins 1990; *Zostrianos* (V111,1) pp.402-403)

[6] (*Nag Hammadi Library,* James M. Robinson general editor Harper Collins 1990, *Zostrianos* (V111,1) p. 430)

[7] (*Nag Hammadi Library,* James M. Robinson general editor Harper Collins 1990; The Teachings of Silvanus (V11,4) p. 387)

[8] (*The Dead Sea Scrolls Uncovered,* Robert Eisenman & Michael Wise, published in Penquin Books 1993, p. 145)

[9] (*The Schocken Book of Jewish Mystical Testimonies,* Louis Jacobs, copyright 1976 by Keter Publishing House, Jerusalem Ltd., The Riders of the Chariot, p. 43)

[10] (A Buddhist *Bible,* edited by Dwight Goddard, Beacon Press, Boston 1970, The Diamond Sutra p. 89)

[11] KJV *Isaiah* 14:9

[12] et al *Proverbs* 15:24

[13] et al *Ezekiel* 31:17

[14] et al *Revelation* 5:3

[15] et al *Revelation* 5:13

[16] et al *Philippians* 2:10

[17] *Pearl of Great Price, Moses* 6:63

[18] *D&C* 88:104

[19] (*The Sumerians,* Samuel Noah Kramer, 1963, The University of Chicago Press, Chicago, pp.123, 132, 134, 135.)

[20] *The Egyptian Book of The Dead,* translated by E.A. Wallis Budge, 1st published in 1967, Dover Publications, Inc. NY.

[21] *The Egyptian Book of The Dead,* translated by E.A. Wallis Budge, 1st published in 1967, Dover Publications, Inc. NY.

[22] (*The Schocken Book of Jewish Mystical Testimonies* compiled by Louis Jacobs, 1976, Keter Publishing House, Jerusalem Ltd., pp.36, 37, 43.)

[23] (et al *Moses* 1:35)

[24] (et al *Moses* 1:37)

[25] (et al *Isaiah* 51:13)

[26] (*The Schocken Book of Jewish Mystical Testimonies* etc. , p. 39)

[27] *(Isaiah* 40:22; italics added)

[28] (Excerpted from *Sefer Yetzirah* by Aryeh Kaplan ©1997 with permission of Red Wheel/Weiser, York Beach, ME and Boston, MA pp.232,233,239)

[29] (*The Amazing Universe*, by Herbert Friedman, published by The National Geographic Society, 1973, p. 172)

[30] (*The Wisdom of the Vedas*, J.C. Chatterji, 1992, The Theosophical Publishing House, Wheaton, Illinois, pp.135, 139)

[31] (*Joseph Smith and the Creation*, William Lee Stokes, 1991, Starstone Publishers, Salt Lake City, p. 66)

[32] (*The Nag Hammadi Library*, James M. Robinson, 1990, Harper Collins Paperback Edition, New York, *The Hypostasis of The Archons*, p. 167)

[33] (et al *Moses* 7:30)

[34] (*The Zohar,* translated by Simon, Sperling and Levertoff, ©1984, London, The Soncino Press, Ltd., 1V, p. 68)

[35] Excerpted from *The Forgotten Books of Eden* ©1927 by Alpha House, Inc. Published by World Publishing, Nashville, TN. Used by permission of the publisher. All rights reserved. , p. 98)

[36] (Excerpted from *The Forgotten Books of Eden* ©1927 by Alpha House, Inc. Published by World Publishing, Nashville, TN. Used by permission of the publisher. All rights reserved. Adam and Eve, p. 7)

[37] (et al Moses 1:29)

[38] (*The Nag Hammadi Library*, James M. Robinson, 1990, Harper Collins Publishers, New York, N.Y., *Zostrianos* (V111,1), p. 408)

[39] (Excerpted from *The Forgotten Books of Eden* ©1927 by Alpha House, Inc. Published by World Publishing, Nashville, TN. Used by permission of the publisher. All rights reserved. p. 82.)

[40] (Excerpted from *Sefer Yetzirah* by Aryeh Kaplan ©1997 with permission of Red Wheel/Weiser, York Beach, ME and Boston, MA pp.185, 187, 188.)

[41] (et al *Zohar* Vol. IV p. 210)

[42] (Brigham Young, October 6, 1854, JD 2:94)

[43] (*Sefer Yetzirah*, et al p. 51)

The Gates of Heaven

The gates of heaven are the doors of entry from our universe to the spiritual or eternal heavens. It is the passageway from the physical dimension to the spiritual fifth dimension, called the dimension of Soul.

In the spiritual dimension, a physical body cannot move at all. Only the soul can move through the spiritual dimension, and it is for this reason that this domain is called Soul. [1]

These entry ways are referenced to in many ways in almost all the sacred scriptures and ancient writings as: Heaven was opened; the heavens were opened; the gates of Heaven were opened; the Heavens were rolled back; the gates of Hell and Hell Gate.

A heaven has many portals of entry from one to another. They could number in the hundreds or thousands as the creation is immense. There are records that describe gate keepers, referred to as watchers, sentinels, and toll-collectors, who monitor the gates and decide whether or not to grant permission to enter that particular

Heaven, depending on the qualification of that soul requesting entry. All scholars, modern and ancient, agree that there appears to be a difference in energy spectra, also referred more esoterically as differences in glory. Matter from one heaven cannot simply pass into another heaven without a correct level of energy or glory. All records agree that anyone who sees a being of higher glory, or is permitted to tour heaven, must first be transfigured. This is a temporary, and apparently authorized change of the mortal body to withstand the glory of the greater being without being burned by its presence. Visits by the resurrected Christ and numerous visits by angels don't seem to require this change. It seems that beings of higher glory may travel to lesser realms, but beings of lower realms must be transfigured to travel into the higher heavens.

The Zohar or *The Book of Splendor*, states: "There are heavens and heavens; to wit, lower heavens with an earth beneath them, and upper heavens also having an earth beneath them. Furthermore, there is in each heaven a chieftain who is in charge of a part of the world and a part of the earth. Each heaven is provided with a certain number of portals, and the charge of each chieftain extends from one portal to the next, and he may not encroach on the sphere of his fellow-chieftain by even so much as a hairbreadth, except he receive authorization to exercise dominion over his neighbor."[2]

The mighty gates of heaven are said to be enormous and soar in height. They can open to accommodate passage of anything, regardless of size, even planets, stars and galaxies. All the gates have names. From some of the ancient writings and scriptures, we have learned the names of a few. In the *Pseudepigrapha, Forgotten Books of Eden*, in *Secrets of Enoch*, there are six sets of gates named the Eastern Gates and six sets of gates named the Western Gates. Other writings give names to some of these gates as the Gates of Righteousness, the Gate of Magdon, which according to Kabbalah is located over the land of Israel, the Gate of Sert, in the underworld, and the Gates of Death. Perhaps, the most infamous of gates in heaven are the Pearly Gates, a nickname humorously applied to the twelve gates of Pearl that open to the holy celestial city, Jerusalem. Though they do not open the heavens, they do open to that city and are note worthy of mention that there are other gates as well in the Heavens. St. Peter is often portrayed as sitting outside these gates, checking a roll or scroll for names of souls permitted to enter. [3] The Gate of Paradise opens into the Garden of Eden.

If the term "Heaven was opened" is used, only one gate is opened from our physical universe to the next spiritual Heaven. If the term, " The Heavens were opened" is used, then all the gates are opened from our physical universe to all the spiritual Heavens.

It is impossible for a mortal being, from this planet, to travel to the spiritual fifth dimension without first under going a change to the body by one of four ways; experiencing death, being transfigured, becoming translated beings, or being resurrected into a perfected body. The physical body we possess is not equipped to endure travel to that realm.

However, an open gate could also mean a portal or worm hole opening into the fourth dimension whereby a mortal physical being, under the right circumstances, could enter. The right circumstances would be entering via spacecraft designed for interdimensional travel, a technology we, as yet, do not possess on this planet. Another circumstance would be a disembodied soul accessing this realm. Lastly, another circumstance would be persons with certain psychic abilities accessing this dimension, whether intentionally or by accident. It would seem that prophets, seers, and remote viewers have the ability to transcend the fourth dimension at least occasionally, and some at will. It is conceivable that advanced beings from other worlds have visited planet earth in the past by traveling in the fourth dimension where time travel is possible. In this dimension it is possible to travel or witness in the present, past, and future time. It would also be possible to travel anywhere in the temporal universe (space) in a short span of time. That is why the fourth dimension is called space-time.

In non-religious writings there are reports of those who have out-of-body experiences (OBE). Parapsychologists generally classify OBE's in two ways. One way is according to the circumstances under which the OBE took place. The basic classification is called an enforced or induced OBE, through an event such as trauma of an accident or emotional shock. The other large class of OBE's happen without any apparent reason---on the verge of sleeping, or sitting in a classroom. These are widely termed natural OBE's. Most of the time these happen spontaneously, but a few individuals can bring on an OBE apparently at will.

Most of the documentation surrounding spontaneous OBE's, including induced OBE's, reveals a common thread. The participant often views his body and the surrounding area from another perspective, usually from above. Reports of disembodied travel to far away places from that point are common, including to the past, the future, and even to the long dark hallway or tunnel with a light at the end. Thousands of participants returned with accurate description of those locations and time periods.

The OBE adept—one who can have an OBE at will—can travel thousands of miles to distant lands or planets, even under the sea or in different time periods, in an instant. [4]

Thousands of well-documented experiments have been conducted to discover the source of these direct observations. The current term for the OBE adept is the remote viewer. Hundreds of thousands of remote viewers have been tested under countless conditions designed to eliminate any chance or error for more than 100 years. There are millions of documented results irrefutable proving the presence of multiple levels of conscious and subconscious existence. OBE'ers have not only accurately described surroundings at impossible distances and time zones, they have communicated with persons and electronic devices at those locations. The proof is irrefutable, and explainable only by the existential travel of the non-physical or spiritual part of the person with the OBE.

The many hundreds of religious writers maintain that a certain level of worthiness is required to pass from one heaven to the next. The records support the state of goodness or righteousness, relative to God, increasing from the lowest heaven to the highest heaven. They all agree that the God of the universes or creator resides in the highest, while the ruler of badness or evil resides in the lowest heaven. There is a common belief that the amount of light or glory increases from the lowest to the highest heavens. In every case where common mortal man was allowed to see God face to face, that person must undergo a temporary physical change or transfiguration to withstand the light or glory of God. The mortal's physical state returns to normal shortly after the visitation is over, but the experience has been recorded as being exhausting. Peter, James, and John underwent this change on the mount of Transfiguration. [5] Joseph Smith saw the Father and the Son in the Spring of 1820, in Palmyra, New York. [6] His testimony of that experience is consistent with those of others who saw and heard the Father. The apostle Paul, Mohammed, Moses, Alma, and other religious figures have also undergone the same temporary physical change. It is clear from these accounts that there is some credence to the elevated glory, light, or energy from one heaven to the next.

Many hundreds of technical writers also maintain that these heavens do exist, and that there is a way for at least the disembodied person to travel between at least two of them. The evidence proves that time may be transcended as well. Precognition not only sees the future, but has empowered many observers to cause a different future by changing the events that led to the observed future. Not stepping onto an airplane, train, or elevator that one has witnessed through remote viewing is going to crash, changes that perceived

74

future by not including the observer in the otherwise unavoidable crash. If the pre-warned person can actually stop the accident from occurring somehow, then the future has been even more profoundly averted.

There is even statistically significant evidence that remote viewers with psycho-kinetic abilities have been able to view the past and change it while in the past, thus producing a comparable set of pasts observable in present time. Again, the evidence is irrefutable and plentiful.

There are places, on this earth, that are alleged to be portals or vortexes of entry to the next Heaven. For these places to be remembered as such, an unusual event occurred at that site linking Heaven and earth. These events were remembered by writings or oral traditions. Such sites are still considered sacred to many. One site is found in the ruins of the ancient lost city of the Incas, Machu Picchu, Peru, located high in the Andes. Situated on a high rise, a large odd shaped rock, called the Intiwatana, actually vibrates continuously. These vibrations or energy force fields can be felt a few inches from the rock without touching it. It is an ancient belief among the Incas living there, that this rock marked the location of an entrance, gate, or portal to heaven and was a sacred site. It is said that there were other rocks, similar to that one, located in some of the other ancient Incan cities that possessed the same properties. The Spanish destroyed these rocks in their conquest of the Inca empire. The one in Machu Picchu is the last surviving rock.

Another site is in Israel. The word Talpiot, which, as the Talmud teaches, is the "hill (tell) to which all mouths (piot) turn." This "hill" is the mount upon which the Temple was built, which Jacob called the "gate of heaven."[7] Jacob saw in a dream, a ladder set up on the earth, and the top of it reached to heaven; and beheld the angels of God ascending and descending on it. He beheld God, who promised him that his seed would be as the dust of the earth and in his seed shall all the families of the earth be blessed. When Jacob awoke from his dream, he declared, "this (place) is none other but the house of God, and this is the gate of heaven."[8]

Later the Jewish temple was built on this site and the temple was viewed as a gate to heaven, until its destruction. The Muslims also consider temple mount a gate to heaven and revere it as a sacred site because the prophet Mohammed is said to have ascended into the seven heavens from that very location. In addition, both Jews and Muslims revere temple mount as the site where Abraham took his son to be offered as a sacrifice, and an angel of the Lord called unto him out of heaven and said, "lay not thine hand upon the lad, neither do thou any thing unto him;".... . And the angel of the Lord called unto Abraham out of heaven the second time and gave him a blessing that his seed would be as the stars of the heaven, and as the sand which is upon the sea shore; and thy seed shall possess the gate of his enemies.[9]

It is believed that the Cairo pyramids, tombs of the Pharoahs, were star gates, launching pads, or ladders to heaven for the souls of these kings to soar upward into the heavens and join in the abode of the gods, by-passing the netherworld. These pyramids have shafts from the tomb, ascending nearly to the top, opening to the outside in view of the constellation of Orion, believed by the Pharoahs to be the place of the abode of the gods.

The following is pure conjecture by the authors that perhaps the mysterious Stonehenge site in England, besides being a calendar or astrological site, was also deemed a gate to heaven. It is well known that the ancient ruin was deliberately built on a ley line, or magnetic energy force field. These lines were known by the ancients in South America, China, Egypt, America, Australia, France, as well as in England. In 1921, Alfred Watkins (1855-1935) re-discovered these energy force fields and concluded that the lines of force were worldwide. Intersecting Ley lines are powerful force points. Ley lines are also referred to by the Chinese as "dragon paths." In Australia they are called "spirit paths." Many circular stone constructions, similar to Stonehenge, and churches were built on these lines in England. In fact, Buckingham Palace sits on a Ley line, as well as two other palaces. The Intiwatana in Machu Picchu is located on ley lines that intersect. In England, the ley lines were anciently referred to as "the old straight ways."

In ancient times circular stone constructions were built on these lines and were considered sacred sites. With the coming of Christianity these circular buildings or monuments were deemed "pagan"; many were destroyed, and later churches were built over these sites. After all, there was no doubt that these were locations of immense power, so why not take advantage of them? Each year, at certain times, the magnetic energy coursing these lines is said to be so strong that the churches and other structures built on the ley lines or dragon paths open their front and back doors to let this mysterious force, called the dragon, pass through. One of the churches, situated on a ley line, is named St. Michaels. The irony here is that the archangel, Michael, was known, in ancient times, as the dragon slayer. Upcoming chapters in this book will discuss more on ley lines or magnetic energy force fields.

The vast spiritual heavens do exist, though invisible to the natural eyes of man. The wind is invisible, yet we know it exists, and feel its effects. We cannot see atoms, protons, or electrons, with the unaided natural eye, yet they exist. Heat and cold likewise are invisible and their effects are known. Almost all-sacred scriptures declare that these universes or heavens do exist; that entry into these heavens is only accessed through one of the many guarded gates. It is impossible, on this side of the veil, to envision the immensity of the creation. It is beyond the scope of mortal comprehension and finite mind of man. Perhaps *1 Corinthians* 2:9 says it best:

> "But as it is written, eye hath not seen, nor ear heard, neither have entered into the heart of man, the things which God hath prepared for them that love him."

End Notes

[1] *(Excerpted from Sefer Yetzirah by Aryeh Kaplan ©1997 with permission of Red Wheel/Weiser, York Beach, ME and Boston, MA p. 145)*

[2] (*The Zohar,* translated by Simon, Sperling and Levertoff, ©1984, London, The Soncino Press, Ltd., Vol. 1V, pp.210, 211)

[3] (*Revelation* 21: 10-21.)

[4] (*"Parapsychology. The Controversial Science,"* by Richard S. Roughton, PhD Ballantine Books July 1991 pp.246-247)

[5] (*St. Matthew* chapter 17:1-9)

[6] (*"Joseph Smith History,"* chapter 1 verse 17, Extracted from *The History of Joseph Smith, The Prophet, "History of the Church,"* Vol. 1, Chapters 1-5. Published by the Church of Jesus Christ of Latter-Day-Saints.)

[7] (Excerpted from *Sefer Yetzirah* by Aryeh Kaplan ©1997 with permission of Red Wheel/Weiser, York Beach, ME and Boston, MA., p. 239)

[8] *(Genesis* 28:17)

[9] *(Genesis* 22:2-18)

The Throne of God

The place where God dwells is referred to as the throne of God. The scriptures all say that it is located in the highest heaven. The highest heaven is the highest spiritual heaven or universe located in another dimension called Soul. Many have mistakenly tried to place the throne of God somewhere in this universe, even in our own Milky Way galaxy. They have overlooked the fact that God lives in an eternal place, not subject to destruction, corruption, or age. Our universe is temporal. Nothing in this universe is eternal. It is a physical dimension, subject to age and corruption. All things in it will eventually perish. *1 Corinthians 15:50*, clearly states this principle:

"Now this I say, brethren, that flesh and blood cannot inherit the kingdom of God; neither doth corruption inherit incorruption."

The following shows us distinctions between eternal and physical heavens in these much quoted verses:

"Lay not up for yourselves treasures upon earth, where moth and rust doth corrupt, and where thieves break through and steal:

But lay up for yourselves treasures in heaven, where neither moth nor rust doth corrupt, and where thieves do not break through nor steal:"[1]

The apostle Peter understood the differences between the eternal and physical nature of men when he spoke of the resurrection as:

"Being born again, not of corruptible seed, but of incorruptible..."[2]

Because we do not know the exact number of spiritual universes, we will simply state that God lives in the highest Heaven or universe, whether that be the third, sixth, seventh or tenth Heaven. For purposes of this book, it is of no consequence.

Abraham, by the power of the Urim and Thummim, saw the spiritual universe in which God dwells and wrote:

"And I saw the stars, that they were very great, and that one of them was nearest unto the throne of God; and there were many great ones which were near unto it;"

"And the Lord said unto me: These are the governing ones; and the name of the great one is Kolob, because it is near unto me, for I am the Lord thy God; I have set this one to govern all those which belong to the same order as that upon which thou standest."[3]

The statement "the same order" further validates that our universe was patterned after the great spiritual universes. Further reading in verse 4 states:

"And the Lord said unto me, by the Urim and Thummim, that Kolob was after the manner of the Lord, according to its times and seasons in the revolutions thereof; that one revolution was a day unto the Lord, after his manner of reckoning, it being one thousand years according to the time appointed unto that whereon thou standest. This is the reckoning of the Lord's time, according to the reckoning of Kolob."

Abraham's writings further say that God lives on a spiritual planet near to the great governing star (sun), Kolob. One revolution of his planet equals 1,000 years of man's time, in our physical dimension, but is as a day to God. Abraham also saw that among all the great governing stars, Kolob is "nearest" or "nigh" to the throne of God; it is "the first creation, nearest to the celestial, or the residence of God."[4]

The knowledge that multiple universes exist and that God lives in the highest Heaven is proclaimed in scriptures and hymns in phases such as: " Glory to God in the highest", "the Highest gave his voice", "the most high God", and "Hosanna in the highest". When we sing Christmas carols using these phases, how many have understood the true meaning of these words?

In 1864, a sermon recorded by Brigham Young said:

"When the earth was formed and brought into existence and man was placed upon it, it was near the throne of our Father in Heaven."[5]

We can now better understand the meaning of *Acts 7:49*:

"Heaven is my throne, and earth is my footstool:"

This marvelous latter day revelation was the missing piece of the puzzle. The doctrine tells that earth was first a spiritual creation and created in the highest spiritual universe, near to the residence of God, so close that it was called his footstool, and that man originally was a spiritual creation placed, in the beginning, on the spirit earth.

Most of the ancient texts refer indirectly to a spiritual creation and fail to make a clear definitive statement. Perhaps, in ancient times, this doctrine was understood by the scribes and deemed not necessary to incorporate repetitious writings that were already accepted as general knowledge. In time, over centuries, the unwritten doctrine probably lost acceptance in the main stream of religious thinking.

Shortly after the Essene movement disappeared, as documented by the *Dead Sea Scrolls*, the history of Gnosticism, as documented in the *Nag Hammadi library*, arose with roots in Judaism, and resurrected once again, the ancient teaching of a dual creation in their writings. In general, Gnostic writings reflect dualism, or a split between self and world, the devaluation of nature, the self-experience of existence, the cosmic solitude of the spirit and the nihilism of mundane norms.[6] The new religion was not well received in the established religious communities of the time. Ancient doctrines, once accepted, now became the writings of heretics. With the conversion of the Roman Empire to Christianity of the more conventional kind, Gnostics were driven underground, hunted and killed for their beliefs, and ultimately eradicated from Christendom. Like the *Dead Sea Scrolls*, the manuscripts of the Gnostics were discovered in earthen jars, in 1945, in the Nag' Hammadi region of Upper Egypt, at the base of a cliff in rock rubble, partially covered by a large piece of a fallen shattered boulder.

It wasn't until 1978, that the manuscripts were finally published in English. For even though the contents of the *Nag Hammadi library* are somewhat sketchy, their manuscripts do reflect the ancient belief of a dual creation, found in Jewish mysticism, long forgotten in the main stream, but clandestinely alive in Kabbalah. For the most part, they were a religious group ahead of their time. Two of their texts in the *Nag Hammadi library* refer to the manuscripts being hidden for safekeeping in a mountain until the end of the times and the eras. Perhaps the finding of the manuscripts in today's times is the fulfillment of that prophecy. Now is the right time to access, understand and rewrite the history of Gnosticism.

The next chapter, *The Dual Creations*, is a compilation of some of the many documented references that support the ancient doctrine.

E. J. Clark & B. Alexander Agnew, PhD

End Notes

[1] *(Matthew 6:19-20)*

[2] (*1ˢᵗ Peter* verses 2-3)

[3] (*Abraham* 3:2-3)

[4] (*Abraham*, Explanation of Facsimile No. 2, Figure 1)

[5] *(Journal of Discourses* XV11, p. 143, talk given July 19, 1864)

[6] (*The Nag Hammadi Library in English*, Revised Edition, James M. Robinson General Editor, 1978, 1988, HarperCollins Publishers, New York, p. 544)

The Dual Creations

The account in *Genesis* is short and clearly implies two creations. *Genesis* states that man was not on the earth.

> "And every plant of the field before it was in the earth, and every herb of the field before it grew: for the Lord God had not caused it to rain upon the earth, and there was not a man to till the ground."[1]

In the *Pearl Of Great Price*, from the book of *Moses*, revelation makes it clear that there were two creations.[2]

> "And every plant of the field before it was in the earth, and every herb of the field before it grew. For I, the Lord God, created all things, of which I have spoken, spiritually, before they were naturally upon the face of the earth. For I, the Lord God, had not caused it to rain upon the face of the earth. And I, the Lord God, had created all the children of men; and

not yet a man to till the ground; for in heaven created
I them; and there were not yet flesh upon the earth,
neither in the water, neither in the air."

Here several important points are revealed. Moses clearly
states that there were two creations, one spiritual and one natural. He
further states clearly that there was no flesh of any kind anywhere.
The scripture also reveals that man was first created in heaven.
In the *Doctrine and Covenants* another account of a dual creation
is found.

"For by the power of my Spirit created I them; yea all
things both spiritual and temporal—

"First spiritual, secondly temporal, which is the beginning of my work; and again, first temporal, and
secondly spiritual, which is the last of my work...."

"Speaking unto you that you may naturally understand; but unto myself my works have no end, neither beginning; but it is given unto you that ye may
understand, because ye have asked of me and are
agreed.

Wherefore, verily I say unto you that all things unto
me are spiritual, and not at any time have I given
unto you a law which was temporal; neither any man,

not the children of men; neither Adam, your father, whom I created."[3]

An account in *The Second Book of Esdras*, taken from the *Apocrypha*, tells of a dual creation.

(6) "And thou leadest him (Adam) into paradise, which thy right hand had planted, before ever the earth came forward."

More concepts of a dual creation are revealed in multiple scriptures from the *Qur'an*. Jonah

1. (10. 4) "To Him is your return, of all (of you); the promise of Allah (made) in truth; surely He begins the creation in the first instance, then He reproduces it, that He may with justice recompense those who believe and do good; and (as for) those who disbelieve, they shall have a drink of hot water and painful punishment because they disbelieved."

2. (10. 34) "Say; Is there any one among your associates who can bring into existence the creation in the first instance, then reproduce it? Say; Allah brings the creation into existence, then He reproduces it; how are you then turned away"?

The Spider

1. (29. 19) "What! Do they not consider how Allah originates the creation, then reproduces it? Surely that is easy to Allah."

2. (29. 20) Say: "Travel in the earth and see how He makes the first creation, then Allah created the latter creation; surely Allah has power over all things."

The Romans

1. (30.11)" Allah originates the creation, then reproduces it, then to Him you shall be brought back."

2. (30. 27) "And He it is Who originates the creation, then reproduces it, and it is easy to Him; and His are the most exalted attributes in the heavens and the earth, and He is the Mighty, the Wise."

The Ant

1. (27.64) "Or, Who originates the creation, then reproduces it and Who gives you sustenance from the heaven and the earth. Is there a god With Allah? Say: Bring your proof if you are truthful."

The concept of a dual creation is also found in the gnostic writings, in *Zostrianos*. [4]

1. "And about this airy-earth, why it has a cosmic model?"

2. (10) "He saw a reflection. In relation to the reflection which he saw in it, he created the world. With a reflection of a reflection he worked at producing the world, and then even the reflection belonging to visible reality was taken from him."

3. P. 430 "I came down to the aeon copies and came down here (130) to the airy-earth. I wrote three tablets (and) left them as knowledge for those who would come after me, the living elect. Then I came down to the perceptible world and put on my image."

From the Gnostic writings of *The Hypostasis of The Archons*:

"...after the pattern of the realms that are above, for by starting from the invisible world the visible world was invented."[5]

From the Gnostic writings of *The Teachings of Silvanus* (V11,4):

"For everything which is visible is a copy of that which is hidden."[6]

In the ancient Kabbalah texts of *The Bahir, The Book of Illumination*, it reads:

"They said to him: It is written (*Lamentations* 2:1), "He threw the beauty of Israel from heaven to the earth." From here we see that it fell. This would indicate that the "land" or "earth" was originally in heaven."[7]

From the ancient Kabbalah texts of *The Zohar, The Book of Splendor*, it reads:

"In the beginning God created the heaven and the earth", means that the lower world was created after the pattern of the upper,......and the latter contains all the varieties of forms and images to be found in the former."[8]

"There are two worlds and they created worlds, one an upper world and one a lower world, one corresponding to the other; one created heaven and earth and the other created heaven and earth."[9]

Another concept of a dual creation is taken from Bahai writings in *Life in the Next World.*

"The difference and distinction between men will naturally become realized after their departure from this mortal world. But this distinction is not in respect to place, but in respect to the soul and conscience. For the kingdom of God is sanctified (or free) from time and place; it is another world and another universe."

These scriptures and writings all testify of a dual creation. To ignore them in the creation accounts is to give an incomplete view of the whole creation story. These accounts will broaden our understanding of the creation in the chapters that are to follow. Even though it might appear to those pragmatists of creationism and evolutionism that these many ancient records are as so many fables and legends, the authors decided that factual evidence is in the eye of the observer. These recorders didn't know each other. They were not contemporaries. Objectively what appears to have happened is that they each observed a series of events, or received a series of revelations or messages.

The authors have tried diligently not to eliminate any of these accounts. Humanity has suffered too many times at the hands of researchers, professors, scribes, and translators. The purpose of this stratagem is to unlock your minds and to let you see the creations with your faculties fully aroused. The greatest minds and leaders of all time have repeatedly expended energy, time, and in some cases their very lives so we could read their accounts of these experiences and these events. The least the authors could do is give the reader the chance to decide for themselves just how, or if the world was created.

E. J. Clark & B. Alexander Agnew, PhD

End Notes

[1] *(KJV Genesis 2:5*
[2] *(Moses* 3:5)
[3] *(D&C* 29:31-34)
[4] (V111,1), p. 406
[5] (*Nag Hammadi Library*, *The Hypostasis of The Archons*, p. 163)
[6] (*Nag Hammadi Library*, *The Teachings of Silvanus* (V11,4) p. 387)
[7] (Excerpted from *The Bahir* by Aryeh Kaplan © 1979 with permission of Red Wheel/Weiser, York Beach, ME and Boston, MA pp.12, p. 109)
[8] *(The Zohar*, translated by Simon, Sperling and Levertoff, ©1984, London, The Soncino Press, Ltd., Vol. 1V, p. 289)
[9] *(The Zohar*, translated by Simon, Sperling and Levertoff, ©1984, London, The Soncino Press, Ltd., Vol. 1, p. 113)

The Spiritual Creation

"In the beginning when God created the Heavens and the Earth, the Earth was formless wasteland, and darkness covered the abyss, while a mighty wind swept over the waters."[1]

"....the earth was without form, and void; and I caused darkness to come up upon the face of the deep; and my Spirit moved upon the face of the water."[2]

"At first there was only Nun, the primal ocean of Chaos that contained the beginnings of everything to come, according to the Egyptian Upper Kingdom creation story."

"Ku, the creator, began to chant over the great watery chaos and as the chant continued Po (the earth) was born"....Polynesian creation story.

"In the beginning was only darkness and water (nammu). Out of Chaos there came many odd creatures:" Sumerian creation story.

"This is Gaia, Gaea: (Earth) the primordial Greek Earth Mother, the first being to emerge from Chaos." Greek creation story.

"What is the Sphere? This is the Womb. It (the Sphere) is the womb of the present, out of which the future is born."[3]

"All was immobility and silence in the darkness, only the creator, the maker, the denominator, the serpent covered with feathers, they who engender, they who create, were on the waters as an ever increasing light. They were surrounded by green and blue."[4] Mayan creation story.

"In order to deceive Darkness, the Savior provokes the creation of the universe from water, part of which transforms into a giant Womb.......and prompts the Womb to produce heaven and earth and all kinds of seed." [5] Gnostic creation story.

"There is also a similarity in the process of creation between the earth and its inhabitants. The earth when created, according to the accounts we have, was covered with a flood of waters....by and by emerging from the waters. This was the birth of creation, the same as we are born here into this world, from one element into another. After having been brought forth from the element of water, the process of creation, or the further development of the earth continued." [6]

The First Day: The Birth of Spirit Earth

Somewhere in a high spiritual universe, in another dimension called the fifth dimension or Soul, near the Throne of God, is a giant parental water planet, whose size dwarfs the physical planet earth. Its vast ocean is called The Sea of Chaos, and often referred to as "the deep" or "the great abyss." According to Webster's dictionary, one definition of abyss is chaos. Many accounts say that high winds sweep across these waters, creating enormous storm tossed waves.

It was within the waters of the Sea of Chaos, likened to a watery womb, that spirit matter was assembled to form the Spirit Earth. The Sea of Chaos is spiritual water bursting with creative energy and unstable elements, producing chaotic or unpredictable matter; matter that bonds then suddenly disintegrates.

Within this watery womb, in "the deep", the embryonic spirit globe was "without form and void", (empty of life forms) "and darkness was upon the face of the deep", meaning that the embryonic waters of the womb were enveloped in darkness while God's Spirit "moved upon the face of the water," organizing and forming the chaotic spiritual elements into stable spirit matter. When one "forms substance out of chaos" one is bringing about a purely spiritual result in the universe. [7]

The Zohar states, "The earth" here is the upper earth, which has no light of its own. It "was" at first in its proper state, but now "void" and without form",...Hence, "the earth was formless and void and darkness and **spirit.**"[8]

97

As all wombs are enveloped in darkness, this suggests that the regions outside the embryonic environment were fully illuminated.

When the gestation period of the spirit globe was accomplished, the spirit earth slowly emerged from its watery womb, in the "deep", and was born into the "light" that God had commanded to shine upon it.

"And I, God said: Let there be light, and there was light. And I, God, saw the light, and that light was good."[9]

Another way of interpretation is that light was already there and God saw that it was good and gave his stamp of approval by stating, "Let the light be."

It is not unreasonable to conclude that the giant water planet orbits one of the giant stars (suns) clustered near the great star, Kolob. This could be the source of the "light" and it was sufficiently directional to illuminate one half of the spirit planet at any one time.

The newly born spirit earth immediately assumes her position, just above the surface of the primordial waters and begins to rotate, experiencing both the light of "day" and the darkness of "night."

"And I God divided the light from the darkness."

"And I, God, called the light Day; and the darkness, I called Night; and this I did by the word of my power, and it was done as I spake."[10] God then moved the newly created planet to another place in space.

"And God said, Let there be a firmament in the midst of the waters, and let it divide the waters from the waters."[11] Now, you probably missed this the first few times you read that. This clearly describes the birthing process. A *firmament* is the mass of stars that make up the heavens. Letting a firmament be in the midst of two planets means one of them moved to another place. Since the Sun and Moon are not in place until a later phase of the creation process, it is evident that this planetary body is under way across the span of space and time.

As with all higher forms of life, the spirit earth has gender. The scriptures and ancient writings indicate that both the spiritual and temporal earths have gender. Ancient writings and beliefs indicate that the spirit earth is male and the temporal earth is female. We are familiar with terms applied to the temporal earth's gender with statements such as: "Mother Earth," she is "the mother of men" who spoke to Enoch (*Moses* 7:48), she "opened **her** mouth" to receive the blood of murdered Abel,[12] and she will have "travailed and brought forth **her** strength" until "she is clothed with the glory of God". [13] Up coming chapters will explain more about the spirit earth's gender and its role in the temporal creation. "And the evening and the morning were the first day."[14]

II Peter 3:8 states, "One day is with the Lord as a thousand years, and a thousand years as one day." This scripture means it takes one thousand years, of physical earth time, for the planet upon which God resides to make one revolution or "day". One day to him is as a thousand years to us on our physical planet earth. The

Lord indicates that it took 1 day (his time) or 1,000 years (our time) to complete the first stage of the spiritual creation. However, the Lord could have used the time frame just to illustrate or point out that his time frame of a "day" was much greater than ours. For the sake of argument we will call 1day (his time) equal to 1,000 years (our time).

The Second Day of the Spiritual Creation

The rotating infant spirit globe was covered with protective embryonic waters which the Lord began to divide and form into a gaseous atmosphere, called a firmament, to surround and contain the spirit earth.

And God said, "Let there be a firmament in the midst of the waters, and let it divide the waters from the waters."[15]

The waters beneath the firmament were those in contact with the parental global surface or The Sea of Chaos. The Lord, having made a protective gaseous atmosphere to surround the spirit globe, then slowly transported and positioned into place his "footstool", leaving the waters below and leaving the waters of the embryonic shroud, above the firmament behind. This division of the waters was not unlike the division of a cell. Dividing water from water was the birth process of a new planet. The *firmament* is also translated as heavens or as an expanse filled with stars. This clearly describes the removal of the spirit earth from its primordial place of birth to its new location, a little nearer to the Throne of God.

"And I, God, made the firmament and divided the waters, yea the great waters under the firmament from the waters which were above the firmament, and it was so even as I spake.

And I, God, called the firmament Heaven."[16]

The gaseous atmosphere of the water planet spirit earth protected its surface water from evaporating into space during its relocation. And, remember that the there was new light commanded by God to appear. This means there was energy and heat provided to this new planet so that it would not freeze into a solid ball of ice as it was transported through space to its new home. After the spirit earth was transported and positioned into place, then God called the firmament Heaven.

"And the evening and the morning were the second day."[17]

The Third Day of the Spiritual Creation

The infant spirit earth was now close enough to the throne of God to symbolically be called his footstool.

And God said, "let the waters under the heaven be gathered together unto one place, and let the dry land appear: and it was so.

And God called the dry land Earth; and the gathering together of the waters called he Seas: and God saw that it was good."[18]

Up to this early point in the infant water planet's history, the ocean that covered the planet was one body of water. When the waters were "gathered together into one place", the first land mass to appear consisted of only one large continent that wrapped around the center of the planet, extending north and southward.

And God said, "Let the earth bring forth grass, the herb yielding seed, and the fruit tree yielding fruit after his kind, whose seed is in itself, upon the earth: and it was so."

"And the earth brought forth grass, and herb yielding seed after his kind, and the tree yielding fruit, whose seed was in itself, after his kind: and God saw that it was good."[19]

It was on this continent that spirit plant life forms first appear and begin to flourish. The newly formed, transplanted, spirit earth is now receiving its light from distant Kolob, in sufficient amounts for plant life to thrive. The planet continues to rotate, experiencing night and day of unknown lengths of time. [20]

"And the evening and the morning were the third day." Keep in mind that although light and darkness had been defined, there was at this time no sun and no moon to rule. It is quite possible that the entire planet was bathed in light, providing unfettered growing conditions for the botanical life on the earth. It can also be imagined that just because God commanded these things to be so, does not necessarily mean that these forms of life were new, but rather that they were placed there under His command.

The Fourth Day of the Spiritual Creation

Once the spirit earth was embellished with plant life, it was necessary to create its own sun, moon and stars, in order for seasons, days and years to be reckoned, and for plant life to continue to flourish. When the spirit sun was created for the spirit earth, its gravitational pull drew the spirit earth into orbit around it. Likewise, the spirit moon, when created, was drawn into orbit by the spirit earth's gravitational pull. The light from this spirit sun was of greater intensity than that received from distant Kolob, enabling the process of creation to continue with higher life forms.

And God said, "Let there be lights in the firmament of the heaven to divide the day from the night; and let them be for signs, and for seasons, and for days and years:

And let them be for lights in the firmament of the heaven to give light upon the earth: and it was so.

And let them be for lights in the firmament of the heaven to give light upon the earth: and it was so.

And God made two great lights; the greater light to rule the day, and the lesser light to rule the night: he made the stars also.

And God set them in the firmament of the heaven to give light upon the earth,

And to rule over the day and over the night, and to divide the light from the darkness: and God saw that it was good."[21]

Many of the ancient writings say that God also created the spirit sun, moon, stars, and all kinds of seeds from the womb in the Sea of Chaos. The *Genesis* account is not entirely silent on this aspect of creation as seen in the next creation day.

"And the evening and the morning were the fourth day."[22] One can clearly realize that the earth is now in a new location. There is a new order for things, with a new star and a new moon as rulers for the heavens. It's a new neighborhood for the newly formed spirit creation.

The Fifth Day of the Spiritual Creation

After the creation of the spirit sun, moon and stars, plant life burst forth in marvelous array and splendor and the whole face of the spirit earth was watered from a mist that arose from the ground. Plant life was flourishing.

"But there went up a mist from the earth, and watered the whole face of the ground."[23] A temperature equilibrium would have existed on an earth that revolved straight up and down on its axis with relation to the sun. There was no winter or fall. The earth had no delta temperature to drive a change in the winds. The mist which occurred as the air reached the dew point at night would fall as silently and completely as the dew upon the grass in a pre-dawn summer.

Then God created spiritual water life, moving creatures and fowls.

And God said, "Let the waters bring forth abundantly the moving creature that hath life, and fowl that may fly above the earth…"

"And God created great whales, and every living creature that moveth, which the waters brought forth abundantly, after their kind,…"[24]

The statement, "Let the waters bring forth," could mean that the primordial spiritual waters of the Sea of Chaos were instrumental in the creation of spiritual aquatic life, moving creatures and fowl.

Then God, the Father planted a garden eastward in Eden and made the garden grow all kinds of trees that produced good fruit. In the middle of the garden grew the tree of life and the tree of knowledge of good and evil. A river watered the garden.

"And the Lord God planted a garden eastward in Eden;[25] And out of the ground made the Lord God to grow every tree that is pleasant to the sight, and good for food; the tree of life also in the midst of the garden, and the tree of knowledge of good and evil.

And a river went out of Eden to water the garden;"[26]

"And the evening and the morning were the fifth day."[27]

The Sixth Day of the Spiritual Creation

The sixth day begins with the creation of Adam then the creation of a host of spirit animal life designed to fill the spirit earth.

"And the Lord God formed man of the dust of the ground, and breathed into his nostrils the breath of life; and man became a living soul.[28] And the Lord God took the man, and put him into the garden of Eden to dress it and to keep it."[29]

"And out of the ground the Lord God formed every beast of the field, and every fowl of the air; and brought them unto Adam to see what he would call them: and whatsoever Adam called every living creature, that was the name thereof.

And Adam gave names to all cattle, and to the fowl of the air, and to every beast of the field;"[30]

After the creation of Adam and the animals, Adam was commanded to name all the animals. Even the oldest zoologists don't know the names of all the animals. Conceivably, Adam could have taken hundreds of years to name and classify all the animals.

And the Lord God said, "It is not good that the man should be alone; I will make him an help meet for him."[31] "And the Lord God caused a deep sleep to fall upon Adam and he slept: and he took one of his ribs, and closed up the flesh instead thereof;

And the rib, which the Lord God had taken from man, made he a woman, and brought her unto the man."

And Adam said, 'This is now bone of my bones, and flesh of my flesh: she shall be called Woman, because she was taken out of man." [32] Adam may have been several hundred or even thousands of years old by the time God made Eve from one of his ribs.

In the beginning man was first created as a spiritual being, out of spiritual dust of the spirit earth. He was created in the image of God, who also is a spiritual being. On the spiritual creation Adam was created *before* the animals, birds, and Eve.

The creation text of *Genesis* gives the account of two creations, howbeit short. Both were created in a six-period time frame. The spiritual creation was created in six days of Gods time and the temporal creation was created over six stages or eras of vast time. However, the spiritual universes were created before the temporal universe and the creation of the spirit earth came before the creation of our temporal earth. The seventh day and seventh era were times of rest for the creators of each creation. Up coming chapters will further clarify the two creations.

"And God saw every thing that he had made, and, behold, it was very good.

And the evening and the morning were the sixth day."[33]

The Seventh Day of the Spiritual Creation

"Thus the heavens and the earth were finished, and all the host of them.

And on the seventh day God ended his work which he had made; and he rested on the seventh day from all his work which he had made.

And God blessed the seventh day, and sanctified it; because that in it he had rested from all his work which God created and made."[34]

According to the *Genesis* account, God, the father, was the architect and creator of the spiritual creation. After completing the spiritual creation, with the works of his hands, he rested.

The spiritual creation was accomplished in 7days of Gods time, and 7,000 years of mans time, as written in the first two chapters of *Genesis*.

E. J. Clark & B. Alexander Agnew, PhD

End Notes

[1] *Genesis 1:1-2 New American Bible*
[2] (*Moses* 2:2)
[3] (Excerpted from *The Bahir* by Aryeh Kaplan © 1979 with permission of Red Wheel/Weiser, York Beach, ME and Boston, MA pp.40, 164)
[4] Excerpted from the *Popol Vuh, The Sacred Book of the Ancient Quiche Maya,* © 1950 by the University of Oklahoma Press. Published by the Publishing Division of the University, Norman, OK. English version by Delia Goetz and Sylvanus G. Morley. Used by permission of the publisher. All rights reserved (Mayan scripture)
[5] *(The Nag Hammadi Library, The paraphrase of Shem* (V11,1), James M. Robinson, General Editor, 1988, Harper San Francisco, pp.338-340)
[6] (Orson Pratt, November 22, 1873, JD 16:314)
[7] (Excerpted from *Sefer Yetzirah* by Aryeh Kaplan ©1997 with permission of Red Wheel/Weiser, York Beach, ME and Boston, MA, p. 134)
[8] (*Zohar* Vol. 1, p. 142)
[9] (*Moses* 2:3-4)
[10] (*Moses* 2:4-5)
[11] (*Genesis* 1:6)
[12] (Moses 5:36)
[13] (D&C 84:101)
[14] (*Genesis* 1:5)
[15] (*Genesis* 1:6)
[16] *(Moses* 2:6-8)
[17] (*Genesis* 1:8)
[18] (*Genesis* 1:9-10)
[19] (*Genesis* 1:11-12)
[20] (*Genesis* 1:13)
[21] (*Genesis* 1:14-18)
[22] (*Genesis* 1:19)
[23] (*Genesis* 2:6)
[24] (*Genesis* 1:20-21)
[25] (*Genesis* 2:8)
[26] (*Genesis* 2:9-10)
[27] (*Genesis* 1:23)
[28] (*Genesis* 2:7)
[29] (*Genesis* 2:15)
[30] (*Genesis* 2:19-20)
[31] (*Genesis* 2:18)
[32] (*Genesis* 2:21-23)
[33] (*Genesis* 1:31)
[34] (*Genesis* 2:1-3)

The Temporal Creation

The temporal creation is the physical or visible creation that is known as our universe. It is the place where our Milky Way galaxy, home to our solar system, is located.

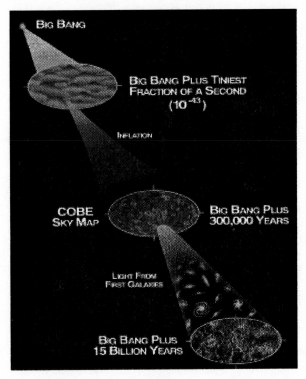

Our universe had a violent beginning. About 13 billion years ago, the universe burst into being through a theoretical event call the Big Bang. The universe itself, and all matter that it contains, seems to have been born in the Mother of all explosions, and its continuing evolution makes abundant use of explosive processes having magnitudes that defy imagination. The Big Bang theory is almost universally accepted by scientists as its beginning. What existed before the Big Bang? Nothing existed before the creation of the universe because time did not exist. Maybe there was matter unorganized and clumped together into one large black hole.

The Big Bang theory states that the universe began with one singular point, which is known as a singularity. All the matter and energy, which is now present in the universe, was squeezed into this singularity of infinitely small volume. From the singularity the universe exploded and released an incredible amount of matter and energy.

From the singularity came not only matter and energy but also space and time. Immediately after the explosion a massive fireball formed called the primordial fireball. This fireball is thought to have been as hot as one million billion billion billion degrees Kelvin. The fireball moved away from the singularity in all directions and from it came elementary particles like neutrons, protons, and electrons.
[1] Since writing this chapter, the Wilkinson Microwave Anisotropy Probe (WMAP), a NASA satellite has been able to take a picture of the images of the Big Bang's afterglow, known as the cosmic microwave background. The remnant glow from the Big Bang

pegs the universe's age to an unprecedented accuracy of 1 percent. Astronomers can now say the universe is 13.7 billion years old. The new data also confirms that the universe began with a brief but humongous growth spurt, dubbed inflation, or the Big Bang. The satellite was also able to delineate the cosmos' composition as 4 percent ordinary matter, 23 percent is invisible stuff called cold dark matter, which prompted the galaxies to coalesce; and 73 percent is so-called dark energy, which has accelerated the rate at which the universe expands. What's more the WMAP satellite revealed that the universe had already made an abundance of stars when it was only 200 million years old, or about one-fifth the age that many astronomers had predicted. This report appeared on *Science News Online,* February 15, 2003, and can be viewed online at http://www.sciencenews.org/20020525/note10.asp.

John N. Bahcall of the Institute for Advanced Study in Princeton, N.J., made the following remarks, "For cosmology, this announcement represents a rite of passage from philosophical uncertainty to precision science. Every astronomer will remember where he or she was when they first heard the WMAP results."

There are two documented sources in ancient writings that support the Big Bang theory. These sources were written about two thousand years ago, long before the Big Bang theory ever surfaced. *The Slavonic Book of Enoch,* in two places (chaps. 11 and 17), speaks about a primordial "great aeon," bearing the inexplicable name of Adoil:

"For before all things were visible, I alone used to go about the invisible things, like the sun from east to west and from west to east…. And I conceived the thought of placing foundations and of creating visible creation. I commanded in the very lowest parts, that visible things should come down from invisible (i.e., the chaos), and Adoil came down, very great, and I beheld him, and Lo! He had a belly of great light. And I said to him: "Become undone, Adoil, and let the visible come out of thee." And he came undone, and a great light came out. And from light, there came forth a great age, and showed all creation which I had thought to create. And I saw that it was good."[2]

Another account of this event is found in *the Secrets of Enoch, Forgotten Books of Eden*, almost word per word as above, but with this additional paragraph:

"And I summoned the very lowest a second time, and said: "Let Archas come forth hard," and he came forth hard from the invisible.

And Archas came forth, hard, heavy, and very red.

And I said: "Be opened, Archas, and let there be born from thee," and he came undone, an age came forth, very great and very dark, bearing the creation of all lower things, and I saw that it was good and said to him:

"Go thou down below, and make thyself firm, and be for a foundation for the lower things, and it happened and he went down and fixed himself, and became the foundation for the lower things, and below the darkness there is nothing else."[3]

According to a primitive Chinese myth, at the beginning of time, the universe had the shape of an egg. When the egg of Chaos broke, a giant, Pan-Ku, came out of it along with two basic elements: Yin and Yang. Yang formed the sky and Yin condensed to become the Earth.

After neutrons, protons, and electrons came from the primordial fireball, these particles spread out across the universe at an incredible rate and soon began to cool down and form elements like hydrogen and helium. These would serve as the foundation and building blocks for the stars and the planets. As the universe continued to cool down, eventually galaxies formed and planets materialized. The universe was a million years old before the first star formed in a galaxy. It has been estimated that there are perhaps 200 billion galaxies or separate star systems extending as far as the Hubble telescope has been able to search.

Galaxies come in different sizes and shapes. Most are found to be spiral-shaped, many to be elliptical, and a few to be irregular. Our own Milky Way galaxy is spiral-shaped. The spirals resemble enormous arms, like a pin-wheel, that spin from a central bulge and coil around it in a thin disk. It is a balanced community of stars at all stages from birth to death. Old red stars fill its bulge and young blue stars, gas and dust define its spiral arms. It is estimated that our Milky Way galaxy contains about 200 billion stars, many of them in clusters of hundreds of thousands.

The diameter of the Milky Way disk is about 100,000 light-years (or 600 quadrillion miles), its central thickness is about

10,000 light-years, and it has five distinct arms. The galaxy needs approximately 240 millions years to complete one orbit of the galactic center . Since its creation the galaxy has made about 20-21 rotations. [4]

Recent discoveries indicate that the galactic center of the Milky Way galaxy contains a giant black hole. The Hubble telescope has further discovered that every galaxy contains a black hole in its galactic center and that black holes may be instrumental in galaxy formation and destruction.

Our solar system is situated within the outer regions of the galaxy, about 28,000 light years from the galactic center. It is situated within a smaller spiral arm, called the local or Orion Arm. Even though it takes the galaxy 240 million years to make a complete rotation, clockwise, the arm that contains our solar system is whipping around 465,000 miles an hour. Our solar system is just a dot on the Milky Way map.

Poet Archibald MacLeish put our earth into perspective: "a small...planet....of a minor star...at the edge of an inconsiderable galaxy in the immeasurable distances of space."

The most accepted theory among scientists is that our solar system condensed from a cloud composed of gas and dust about 4. 5 billion years ago. The process produced heat that ignited the sun and formed young molten planets around it. Later the planets cooled into rocky cores that became the foundation blocks of building into larger planets by a process called accretion-they accumulated dust and rocks, by violent impact bombardment, to eventually become planets. This process took about 3 million years.

Psalms 102:25 describes the early event of earth and its solar system:

"Of old hast thou laid the foundation of the earth, and the heavens are the work of thy Hands."

The accretion process was described years ago by Joseph Smith who revealed that the physical earth was constructed from the destroyed remains of other creations that had not filled the measure of their creation.

"This earth was organized or formed out of other planets which were broken up and remodeled and made into the one on which we live."[5]

Early in the formative period of the hot molten earth, a very large object (as big as Mars or more) collided with the earth and ejected material that cooled and formed the earth's large moon. Detailed information from the Moon rocks led to the now widely accepted impact theory. The Moon, once formed, continued to accrete material to it, by big impacts, possibly adding rock with a composition different from the rest of the Moon, accounting for some unexplained lunar interior features.

In 1974, Dr. William Ward, a scientist, discovered that the presence of our large moon acts to stabilize the variation of the earth's rotational axis. Without this stabilization, extreme climatic changes would occur that would render the planet uninhabitable.

115

It seems that having a large moon may be one of the key factors necessary for a habitable earth-like planet.

After several million years, the impact rate declined, Earth's magma oceans cooled and a stable crust formed. Later water began to leak out of the earth to form oceans and the atmosphere, further cooling the earth sufficiently for life forms to originate.

Genesis is a narrative of the creation process, that supports the above creation theory as follows: chaotic matter not only preceded the planets creation; it became the planet Earth. After the Earth was brought forth into existence, it produced water. And when water, in liquid form, could exist, it (water) gave rise to life. [6] In the infancy of the planet, the oceans were shallow and easily warmed. When land began to merge, the warm waters kept land temperatures tropical. About 250 million years ago the continents rose and collided to eventually form one super-continent called Pangea. Geologists have mapped out the sequence of geological time periods as shown in the following table.

Geological Eras, Periods and Epochs

Millions of Years	Era	Period
2 12 29 41 51 60	Caenozoic (Recent life)	Quaternary Tertiary (58,000,000 years)
130 160 185	Mesozoic (Middle life)	Cretaceous (70,000,000 years) Jurassic (30,000,000) Triassic (25,000,000)
210 265 320 360 440 540	Palaeozoic (Ancient life)	Permian (25,000,000 years) Carboniferous (55,000,000 years) Devonian (50-56,000,000 years) Silurian (40,000,000 years) Ordovician (80,000,000) Cambrian (100,000,000 years)
At least 840	Proterozoic (Primitive life)	Sometimes divided into 300,000,000 years or longer
unknown	Archaeozoic (Beginning of life)	Duration unknown

Conventional Tertiary and Quaternary Periods

Period	Epoch			Millions of Years
Quaternary	Holocene			. 011
	Pleistocene	Upper	Ice Ages	2
		Middle		
		Lower		
Tertiary	Pliocene (10,000,000 years)			12
	Miocene (17,000,000 years)			29
	Oligocene (12,000,000 years)			41
	Eocene (10,000,000 years)			51
	Palaeocene (9,000,000 years)			60

Revised Chronology for the Tertiary and Quaternary Periods in light of the cataclysm. (See chapter on *The Cataclysm*)

Eras	Periods	Epochs	Stages	General Conditions	Approx Dates
Caenozoic	Quaternary	Holocene (Later)	Historic	Present condition	2,300 – today
			Sub-Atlantic	Dry mild	2,700-2,300
			Atlantic	Wet cold	3,450-2,700
			Boreal	Dry cold winters Warm summers	5,500-4,900
		Holocene (Earlier)		Glaciers melt	
			Pre-Boreal	Dry cold	8,000-7,500
			Sub-Arctic	Wet cold	11,400
			Pleistocene	Severe, rapid development of glaciers	11,500
	Cataclysm				
	Tertiary	Pliocene	Equable		14,000,000
		Miocene	Equable		29,000,000

Eras are divided into several Periods and each Period is divided into Epochs. Epochs are subdivided into even smaller divisions.

Animals with hard shells appeared in great numbers for the first time during the Cambrian period. The Early Devonian period was the age of fishes. Plants took over the land and became so abundant that the first coal deposits formed in the tropical swamps.

By the Late Permian period, reptiles had spread across the face of the super-continent of Pangea. Ninety-nine percent of all life perished during the extinction event that marked the end of the Paleozoic Era.

After the permo-triassic extinction, life began to rediversify. The super-continent, Pangea, began to rift apart and by the Cretaceous period the continents were separated.

Mesozoic means "middle life". This era is divided into three periods: Triassic, Jurassic, and Cretaceous. Almost all the modern orders of insects appear during the Mesozoic periods.

During the early Jurassic period, the dinosaurs spread across Pangea. About the middle of the Mesozoic Era, during the Jurassic Period, at the Chicxulub impact site, a 10 mile wide comet caused global climate changes that killed the dinosaurs and many other forms of life.

Close to the middle of the Miocene Epoch, the world assumes a modern configuration of land mass. The ice ages occurred during the upper and middle Pleistocene Epoch. Today we now live in the Holocene Epoch, a modern day era. Geologists indicate that we are entering a new phase of continental collision that will ultimately result in the formation of a new Pangea Super-continent. Global climate is warming which will cause global climatic changes in weather patterns affecting plant and animal life.[7]

This is a broad overview of our planet's creation history and future physical and climatic changes. Those students of earth science who desire a more detailed description of the geological periods may find the information in university level text books. The temporal creation is millions of years old and was brought forth over periods of time. In the order of time, and in the succession of events, the temporal creation of the heavens and earth, and all things contained therein, differs from the spiritual creation.

The Lord told Abraham that the earth upon which he stood, the physical earth's creation or preparatory epoch consisted of six "times" or periods to bring it forth and eventually prepare its surface to sustain modern plant and animal life, including man.[8] The *Qur'an* states in *Surah* V11 54: "Surely your Lord is Allah, who created the heavens and the earth in six periods of time, and He is firm in power."

The Sefer Yetzirah or *The Book of Creation* speaks of the six cycles of creation that occurred over vast periods of time.[9]

This is further validated in *Fenton's The Holy Bible in Modern English* , with the following first and fifth verses of *Genesis*:

"By Periods God created that which produced the Solar systems; then that which produced the Earth."

"And to the light God gave the name of Day, and to the darkness He gave the name of Night. This was the close and the dawn of the first age." (Note: Each creative period was called an age.)

The scriptures and ancient writings indicate that God, the son, along with other noble beings, acting under the direction of the Father, were the creators of the temporal universe, the galaxies, our solar system and the planet which we now live.[10]

"For we saw him, even on the right hand of God: and we heard the voice bearing record that he is the Only Begotten of the Father---

That by him, and through him, and of him, the worlds are and were created, and the inhabitants thereof are begotten sons and daughters unto God."[11]

"God, who at sundry times and in divers manners spake in time past unto the fathers by the prophets,

Hath in these last days spoken unto us by his Son, whom he hath appointed heir of all things, by whom also he made the worlds;"[12]

"And worlds without number have I created; and I also created them for mine own purpose; and by the Son I created them, which is mine Only Begotten "[13]

That the spiritual creation and the temporal creation had different creators is attested by the following writing found in *The Zohar* previously quoted in the chapter, *The Dual Creations* as follows:

"There are two worlds, and they created worlds, one an upper world, and one a lower world. One corresponding to the other; one created heaven and earth (the Spiritual Creation), and the other created heaven and earth (The Temporal Creation)." Words in parenthesis added.

After the physical earth was formed and the many elements organized, God simply let the earth "bring forth", when the conditions are right, without further intervention. This is clearly demonstrated in the following scriptures:

"And God said, Let the earth bring forth grass"[14]

and 'let the waters bring forth abundantly the moving creatures that hath life"[15]

"And God said, Let the earth bring forth the living creature after his kind, cattle, and creeping thing, and beast of the earth after his kind: and it was so." (*Genesis* 1:24)

It appears, from the above scriptures that earth and water in themselves have the capability to "bring forth" living things when conditions are right. The process of "bringing forth" was not instantaneous, but occurred over a vast period of time, and in that respect the *Genesis* account of creation supports evolution. By the same token, this capability to seemingly "bring forth" is a power that has been erroneously labeled by man the evolutionary laws of nature and the powers of nature. These laws of nature are the manifestation of the powers of God. Simply stated; God reveals himself through nature.

The following is excerpted material taken from, *The Seer, Powers of Nature.*

"The all-pervading, omnipresent substance is the Holy Spirit (or Life Force) existing in inexhaustible quantities, and extending through the immensity of space. It is the light, and the life, and the power of all things."

"To search out the laws of nature is nothing less than searching out the laws by which the Spirit in nature operates. Man is continually beholding these wonderful operations, but because he does not behold the acting agent, he ascribes the effects to blind, unintelligent and unconscious matter: as well might he ascribe the attributes of the divinity to a wooden idol. The light shines all around us, and is manifested in an infinite variety of wise and beneficial results, but so great is the darkness of man, that he perceives not the light; or as our great Redeemer has said, "The light shineth in darkness, but the darkness comprehendeth it not."

"Man is continually experimenting with the powers of nature, but he perceives not that those very powers with which he is so familiar, are nothing less than the manifestations of the power of God through the elements which are His tabernacle. It is God, or in other words, the Holy Spirit (or Life Force) which is associated in a greater or less degree with every particle of matter in the universe. It is this holy and All-wise substance that is omnipresent, pervading universal nature, governing and controlling worlds without number, producing and superintending the grand and august movements of the combined whole, as it stretches itself out on every side to infinity. It is this All-wise, Omnipresent, and Almighty substance, that unites system with system, under its own forces, so regulated, as to maintain an eternal bond of union, and yet so nicely adjusted as to prevent worlds from rushing on worlds, as they fly with inconceivable velocity in their appointed orbits."

"Were it not for the presence of this all-pervading substance, matter would be wholly devoid of force. The great central force of gravitation could not exist: matter could not have been collected into worlds; or if collected, there could have been no adherence of its particles—no chemical combinations—no formation of solids or liquids—no organizations of any kind—no varieties of matter—no hard and impenetrable atoms; but all substance, without force, would have been infinitely divisible, without properties of any kind, except the property of existence in space. Without this All-wise substance as the acting agent for stability in the laws of nature, this universe would cease to exist."[16] Note: Words in parenthesis added.

The great Architect of the temporal creation does intervene from time to time in the process of creation in order to bring about the desired results. Perhaps this is why mass extinctions occurred, in the past, in order that new forms of life could be introduced to prepare the planet for that particular time period of creation.

Once life forms were "brought forth" they were given the ability to reproduce after their kind. General characteristics of a "kind" can be altered by climate, and environment but science has produced no evidence that at any time in the course of the development of life on earth has one kind of creature developed into another kind of creature.

These evolutionary laws of nature are nothing less than the power of God, acting according to prescribed laws, fixed and unchangeable in their mode of operation, and changes only when the authority that gave them directs a deviation.

Ancient writings and scriptures attest that some things were also created and did not evolve. Early plant forms, mammals, reptiles, and fish did evolve and some of their descendants are on earth today but there is also a population of modern day animals and modern day birds that were created and did not evolve. This will be explained in the chapter, *The Arrival of Noah*. The age old question of whether man was created in the image of God or did he descend from primates will also be addressed in the chapter, *The Lost Civilization of Mu*. To do so now would be getting the cart before the horse.

Finally, we as earthlings are all "wrapped up" in our own little earth world. It is the "center of focus" in our life. But, if you put the creation of this earth into proper prospective you will quickly discover that it is probably neither the biggest or most important work of the creator in this universe. It is evident that the earth is only a very small portion of the temporal universe. It wouldn't exist were it not for our solar system in the Milky Way Galaxy, one of billions of galaxies in an immense universe...so immense that we have yet to find its limits.

End Notes

[1] (*http://libraryquest. org/28327/main/universe/cosmology/big_bang. html)(viewed April 26, 2002)*

[2] From **ON THE MYSTICAL SHAPE OF THE GODHEAD** by Gershom Scholem, English translation copyright ©1991 by Schocken Books. Used by permission of Schocken Books, a division of Random House, Inc. pp.98-99)

[3] (Excerpted from *The Forgotten Books of Eden* ©1927 by Alpha House, Inc. Published by World Publishing, Nashville, TN. Used by permission of the publisher. All rights reserved. chapter XXV1, p. 90)

[4] (*The Amazing Universe* by Herbert Friedman, PhD. , The National Geographic Society, 1975, p. 134.)

[5] (Joseph Smith, January 5, 1841, Words p. 60)

[6] (*Joseph Smith and the Creation*, William Lee Stokes, Starstone Publishers, Salt Lake, Utah, 1991, p. 40)

[7] (www. scotese. com) (viewed 12/2/00)

[8] (*Abraham* 4:11-12, 20-22, 24-25)

[9] (Excerpted from *Sefer Yetzirah* by Aryeh Kaplan ©1997 with permission of Red Wheel/Weiser, York Beach, ME and Boston, MA, p. 187)

[10] (*Abraham* 3:24) (*The Zohar VI* p. 113)

[11] (*D&C* 76:24-25)

[12] (*Hebrews* 1:1-2)

[13] (*Moses* 1:33)

[14] (*Genesis* 1:11)

[15] (*Genesis* 1:20)

[16] (*The Seer, Powers of Nature*, March 1854, Orson Pratt)

The Interaction of Universes

The great universes both spiritual and temporal are referred to in ancient writings and scriptures as the invisible and visible creations.

As previously stated in the chapter, *The Eternal Heavens*, these great universes, both spiritual and temporal, are layered one above the other and rotate around a central axis, called the Teli.

These great universes occupy a small amount of endless space and are so constructed to allow for expansion in all directions. As our universe (the visible creation) expands in all directions, so do the invisible universes. The authors envision the multiple layered universes similar to inflating a balloon that has pictures on it. The more the balloon is inflated, the more the pictures expand in all directions but yet remain in their same position on the balloon, only larger in size. This stretching out or ballooning can continue forever as the space that surrounds these immense universes is endless.

Incredible as it may seem, there is the possibility that other stacked up universes exist elsewhere in the realms of endless space. Envision them like galaxies found in our universe, only drifting in endless space also referred to as the *outer darkness*. As there are billions of galaxies in our universe, so there may also be an infinite number of stacked up balloon universes floating in endless space, separated by great gulfs of immeasurable distances.

If God, the Father, created the spiritual universes, and God, the Son, created our own temporal universe, (in this multiple layered balloon universe) would it not be unreasonable to assume the possibility that other stacked up, multiple layered, balloon universes, exist elsewhere and are the creations of other creator gods? If these other creations are patterned one after another, then they would closely resemble our own stacked up multiple universes. If you could leave the boundary of our temporal universe and travel into the outer darkness, one could possibly see many other stacked up balloon universes which would appear to be like galaxies in our temporal universe.

Multiple stacked balloon universes floating in the outer darkness of endless space as illustrated by EJ Clark.

Revelations in ancient and modern scriptures attest to the plurality of gods. One modern day doctrine on the creation of the worlds states:

"In the beginning, the head God called together the Gods and sat in grand council to bring forth the world. The grand councilors sat at the head in yonder heavens and contemplated the creation of the worlds, which were created at the time."[1]

The apostle Paul declared there are Gods many and Lords many, but to us there is but one God and we are to be in subjection to that one.[2]

It is evident from the above teachings that God, the Father, was appointed by a grand council of Gods, in the beginning, to be the creator God of a spiritual and temporal creation and to be our God.

As previously written in the chapter, *The Eternal Heavens*, the great universes are separated by a veil like structure, often described as curtains which are invisible to the natural eye of man. The veils act as a filter, channeling down from the highest heaven the correct amount of energy required in each lower universe to keep them in sync. Without the filtering veils, the energy is so strong that it could destroy universes.

The *Hebrew Book of the Dead* states, "Gershom Scholem wrote: Between the World of Atziluth (highest universe) and that of Beriah (universe under Atziluth) and similarly between each of the following ones there is a curtain or partition wall…the power which emanates from the substance passes through the filter of the curtain. This power then becomes the substance of the next world of which again only the power passes into the third and so through all four spheres (universes)."[3]

The Vedic texts call this energy Akasha, Ethereal Space, and describe it as a webbing of straight and curved lines of living energy that permeates each universe. These Lines of Force spread everywhere and weave themselves like webbing of something woven into the Ether or Ethereal Space. The theories of physics tend to do away with the conception of the ether as a "jelly-like" substance, but science may find that there may still be something like the ether, but made up of lines of force, as the Vedic conception has it. [4] A facsimile from the *Book of Abraham* depicts this energy as follows in Fig. 5. "Is called in Egyptian, Enish-go-on-dosh; this is one of the governing planets also, and is said by the Egyptians to be the Sun, and to borrow its light from Kolob through the medium of Kae-e-vanrash, which is the governing power which governs fifteen other fixed planets or stars, as also Floeese or the Moon. This planet receives its power through the medium of Kli-flos-is-es, or Hah-ko-kau-beam. The stars, represented by numbers 22 and 23, are shown receiving light from the revolutions of Kolob. Fig. 2. Stands next to Kolob, called by the Egyptians, Oliblish, which is the next grand governing creation near to the celestial or the place where God resides; holding the key of power also, pertaining to other planets."

Pearl of Great Price; Book of Abraham
Facsimile #2.

This "borrowing of light" is the channeling of energy from other dimensions or spiritual universes, down into our temporal universe along lines of force or mediums described above. The source of this energy is from the revolutions of the giant stars, Kolob and Oliblish, in one of the high spiritual heavens, near to the throne of God. Are Kolob and Oliblish rotating stars that generate energy much like a generator? Kolob is a giant star of immense size, probably larger than our solar system and maybe larger than the largest star ever created. Oliblish is apparently of equal size. The exact mechanism of this power source is not yet understood but clearly the ancients understood its principal effects on the spiritual and temporal universes.

Most all ancient cultures adopted a name for the etheric energy substance that traveled through the dimensions in spirals and was called the etheric "water" of the cosmic ocean, which was also called "Astral fluid," by medieval magicians, "Fiery water," by Kabbalists, and also referred to as the "Life Force," "Spiraling Life Force," "Serpent fire," and "Solar Spirit." In Polynesia it was called "Manna." The Hindus refer to this substance as "Prana" and the Chinese refer to it as "Chi"(pronounced Ki by the Japanese). It was known to the Egyptians as "Ka". The American Indians referred to the Life Force as Orenda (Iroquois), Wakan (Sioux), and Manitou (Algonquins). To the Mayans it was the "Plumbed Serpent." The Hebrews called it the "water of life," and Christians call it the "Holy Spirit." Today's scientists approximate it with electro-magnetic energy, which is actually a densification of the "Life Force" but similarly travels in spirals.

The ancients called the creator of the universe, the Primal Serpent who had a whirlpool (chaos) or spiral shape and was symbolized by the snake motif because the snake's body moves and coils in spirals. Its sacred number was the number seven because the serpent had seven principles or seven coils in its body. Christians today could equate the serpent's seven principles to the Creators seven spirits spoken of in the *Bible, Revelation* 1:4. The seven coils or the seven parts of the Dragon formed the Seven Governors (governing seven stars of the Pleiades) that control the world by a mysterious force called Destiny. (basis of astrology?) The word Serpent contains seven letters, which in numerology all the letters in the word reduce to the number seven and the word pivots around the "p" which has its own value of seven. In our part of the galaxy, the seven stars of the Pleiades, represented a manifestation of the Primal Serpent Creator. The Greek Gnostics referred to the Pleiades star group as the "seven pillars" of Sofia, which was an embodiment of the "wisdom of God."[5] In the *Bible, Job* 38:31 states:

"Canst thou bind the sweet influences of Pleiades, or loose the bands of Orion."

It is quite possible that the gravitational waves generated by these immense bodies moving or revolving through space may be one of the sources of this energy. Many of the nearest black holes and large stars have been mapped using the Laser Interferometer Gravitational Wave Observatories (LIGO) in the United States. The direction and magnitude of these waves are measured as the periods of the waves are gathered. Early ground potential measurements—the amount of voltage a specific compound or atomic species contains at ground state—recorded some of the effects of gravitational waves. Some of these compounds or pure species emitted a small amount of converted energy each time the gravitational wave passed through them. Converting gravitational into small amounts of electrical or light energy was very interesting and had many applications that are still being explored. For the author's purpose, it is remarkable to observe that energy sent out from these large bodies—most notable for our subject are Kolob and Oliblish—can be converted into another form and quite possibly fuel a small reaction billions of light years away.

The ancients believed that the seven stars played a role in the creation and continue to influence our planet. They become visible in the springtime and signal the time to plant. The stars in Orion herald the arrival of winter when they first become visible. The bands of winter are loosed when spring arrives and life is renewed. The following writing is taken from the book, *THE GOLDEN AGE*, July 13, 1927, p. 650, and states the following:

> "….a new theory has just been advanced that the so-called Millikan or cosmic rays, which are of uniden-tified origin but are believed to come from the sun and other stars, furnish the energy of the living cells.

It may be that this is correct and that the Millikan rays are emanations of divine power, the "sweet influences of the Pleiades" of which God spoke to the Prophet Job."

The cosmic rays were called the "Life Force" and "fire serpents" that supplied life giving energy to the earth. The ancients believed that the main source of this preserving energy came from our sun and the Pleiades and were worshipped as a manifestation of the creators' power. The Egyptian and Mayan astronomers were obsessed with sun spot activity and fertility. Perhaps they were right in the knowledge that the cosmic rays furnish the energy of living cells.

It was common knowledge among the ancients that the "Life Force" entered the Earth at certain energy conductive points called "vortexes" or "dragon lairs." These collecting points then dispersed the energy around the planet throughout a "world grid" network of energy channels called dragon lines, dragon paths, straight paths, spirit paths, or ley lines. [6] Ley lines or straight paths can be seen across the landscapes in Bolivia, at Nazca in Peru, as well as in France, England, Ireland, Australia, and China. The energy channels frequently followed rivers and underground streams. When two or more lines intersect, the point of intersection was called a dragon's lair. There is a high level of energy in a dragon's lair. The energy forms a pattern of seven concentric circles that spiral up and down and reflect the Serpent's seven principles. These seven principles are manifested during the spring and fall equinoxes at the Temple of Quetzalcoatl, the Plumed Serpent, at Chichen Itzá, Mexico. During the equinox the shadow of a serpent appears, composed of seven triangles (the seven principles) that represent the seven coils or curves of the serpent's body, which slithers down the steps of the Mayan temple and unites with a stone serpent head at the foot of the steps. The Earth's subtle grid of lines and vortexes is also known as its "Dragon Body."[7]

Dragon lairs were marked to facilitate the movement of the "Life Force" in and out of the Earth. The ancient people placed different types of conduits over these vortexes to aid in the dispersion of the energy force and to monitor its circulation and keep it at a safe level. They were able to siphon off high levels of excess energy preventing the rise of earthquakes and volcanoes. The vortex facilitators took the form of earthen mounds, standing stones, stone circles, and pyramids and were considered sacred sites. The word pyramid means "fire in the middle."[8] One of the sacred names of the Egyptian Sphinx, is "Neb", which means 'the spiraling force of the universe.'[9]

At Heliopolis, in Egypt, where the Sun was considered to be the main source of the "Life Force", the solar orb was worshipped as a heavenly manifestation of the creator and the solar rays were recognized to be needles or shafts of light which emanated from the Solar Spirit, which they named Ra. These solar rays were artfully rendered as slender, tapering stone columns called obelisks, "needles," and "frozen snakes," The first obelisk in Egypt was built over a dragon's lair site called On, or Anu Heliopolis, known as the sacred "City of the Solar Spirit." Pillared temples were also placed over dragon lairs to amplify and ease the movement of the serpent or "Life Force" in or out of a vortex. Each pillar of the temple represented a cosmic tree upon which the serpent or Life Force continually moved between Heaven (the roof of the temple) and Earth (floor of the temple). A dragon's lair was usually identified with inscribed recurring spirals somewhere upon the site or carved on temple blocks. In Mesoamerica, the Sun was recognized to be the primary source of the "Life Force" and the natives created solar icons as an orb projecting seven rays that represented the seven principles of the creator, the Pleiades.

The Moon was considered a secondary source of the "Life Force" especially during the waxing and full Moon period. Different gods were associated with the Moon and its rays. In Egypt it was the gods Seth and Thoth-Hermes. In India it was the goddesses Shakti and Kali.[10]

It is interesting to note that the "spiraling Life Force" is reflected at the most minute level of the human body in the DNA. The spiral shape is the basis of the double helix configuration of the DNA molecule.[11]

Another interesting fact is that most of the galaxies in our universe are rotating spiraling galaxies whose form resembles enormous whirlpools.

Certain people called "dowsers" who locate water for wells by "witching" with a forked stick are tapping into the energy force field, which always is strong over water and even stronger in a dragon's lair. The wood of a peach or willow tree bends toward water which energy force field attracts like a magnet. Metal coat hangers are also used to "witch" water in a similar fashion.

The magicians of ancient Egypt through the practice of magic, either white or black, were able to control the universal "Life Force" or "astral light" by manipulation of the essence (astral fluid, energy force field) to produce the desired phenomena. This is why they were so powerful and feared. Nebuchadnezzar, the king of Babylon, was accompanied by his chief magician, Nergal-Sharezer, into a battle against King Zedekiah. When the King of Judah saw the magician dressed in his magician's robes, surrounded by a great band of warriors, it is said according to Biblical scripture that the Jewish king fled with great haste in the opposite direction.[12]

One tribe of Mayas called their primordial serpent creator, Huracan, the serpent with the whirlpool body and shape. From Huracan evolved the word Hurricane. Modern physics asserts that the primeval form of matter or energy is shaped like that of a geometric whirlpool or spiral, something the ancients knew all along.[13]

Adolph Hitler had the mind of an inquiring mystic. He was well educated in the ancient sciences and even adopted the swastika, a whirling cross that symbolizes the spiraling LIFE FORCE of the universe, to be the symbol of Nazi Germany.

Just recently, two of NASA's Great Observatories, the Chandra X-ray Observatory and the Hubble Space Telescope captured in pictures the spectacle of matter and antimatter propelled to near the speed of light by the Crab pulsar, a rapidly rotating neutron star, the size of Manhattan, in the Crab Nebula, that acts as a cosmic generator! This new and exciting discovery was released to the press in September 19, 2002, after the first part of this chapter was written. This confirms the existence of rapidly spinning generator stars and the possibility that the giant stars, Kolob and Oliblish, are forms of these types of power supplying stars. By understanding the Crab pulsar, astronomers hope to unlock the secrets of how similar objects across the Universe are powered and how they influence other heavenly bodies.

The ancient world knew how to harness the "Life Force" electric energies. It was used to provide a strange form of illumination, to lighten heavy objects, and to separate inertia from gravity, so objects would hang in the air without physical support. Such methods were used to move many of the megalithic stones used in construction of the megalithic cities, cyclopean walls, and many sacred sites such as Stonehenge and even the Great Pyramid of Egypt, an engineering feat that would be difficult even with today's technology.[14] Some of the massive blocks used in the construction of a canal in Puma Punka, Bolivia weigh as much as 300 tons and are up to twenty-seven feet long. Do we have a crane today that could lift it?

Proponents of the "world grid" also maintain that major ancient megalithic cities and sites such as Stonehenge, pyramids and other ancient structures are deliberately located at key points on the grid. Why? Simply to utilize the earth's natural "world grid" energy supply, knowledge once known to ancient civilizations, now lost, but maybe not forever.

Below is a mural found in the Temple of Hathor, Egypt, that seems to depict the collecting of the "Life Force" energy from the ground. The instruments have a serpent image on the sides that indicate that these are collectors of that type of energy. Both instruments have a serpent tail connecting to a common storage unit with a Pharaoh or god (Osiris?) sitting on the top with a solar disc on his head. This could be a type of battery. Note, each collector is slightly different. The serpents have one head raised and one head looking back. The collector on the left has a little different addition on its pole leading to the ground as the third concentric ring from the bottom has connectors that attach to the main collector. The collector on the right doesn't have this. (male and female or positive and negative energy?)

Djed columns were used throughout Egyptian temples like Abydos and Dendera and were apparently lit by electric lights. It appears that djed columns were often used as lights in temple activities. The type of connections depicted in numerous Egyptian reliefs and drawings leads one to surmise that direct current was utilized. However, many researchers claim that the devices could have utilized alternating current as well. It is interesting to note that the above illustration shows the djed columns supporting large energy collectors. Although the actual workings appear to by symbolic in nature, the fact that the djed columns and the long braided cords were used together clearly indicates either an illuminating device or an energy device of some type. These type of ringed columns were used to adorn walking staffs, building columns, and appear to all be constructed with the same motif, as in the illustration below:

Djed column

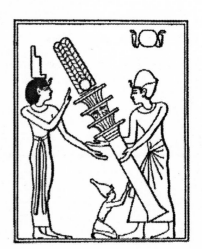

Portable djed columns

Portable djed columns may have been used only in the wealthiest of Egyptian homes or public buildings. Ordinary peasants burned oil lamps for light. When considering the *Edison Effect* of the electric light on modern society, one can only image the effect artificial light had on the Egyptian community. Before the rediscovery of artificial light in modern America, society generally went to bed at dark and arose at first light. Ours is definitely a round-the-clock world with almost universal use of the electric light.

Edward Leedskalnin, an immigrant to America, once said, "I have discovered the secrets of the pyramids, and have found out how the Egyptians and the ancient builders in Peru, Yucatan, and Asia, with only primitive tools, raised and set blocks of stone weighing many tons!" A small statured man, barely weighing 100 pounds, single-handedly devised a means to lift and maneuver blocks of coral weighing up to 30 tons each to construct a complex known today as the Coral Castle, in Florida. He always worked at night and never let anyone observe how he moved the huge blocks of coral that on average the weight of a single block used in the castle was greater than those used to build the Great Pyramid. He did it without electricity and with simple tools employing the knowledge he found in books on magnetic current and cosmic forces. If that wasn't enough, in 1936, when the area around his property became too populated, he decided to move his creation to a new location near Homestead, on a 10-acre parcel of land that he purchased. Borrowing a neighbor's tractor, he worked several months, loading a homemade trailer at night with pieces of his castle and moved them to his new location by day. He moved hundreds of tons of rock alone and without benefit of mechanical equipment, refusing to let anyone observe him while at work. He literally moved a mountain by himself, a feat unequalled in all history!

He took issue with modern sciences understanding of nature and flatly stated that science is wrong about electricity. His concept of nature is simple. All matter consists of individual magnets, like atoms, and it is the movement of these magnets within materials and through space that produces measurable phenomena, such as magnetism and electricity. (the "Life Force"?) He claimed science only knew one side of electricity, that it didn't have electrons. The other half of this force is magnetism. He claimed to have found the lost knowledge in a library in Europe. The following are some excerpts taken from his book, titled, *Book on Magnetic Current Part 1*:

"Voltmeters and ampere meters are one-sided. They only show what is called by instruction books, positive electricity, but they never show negative electricity. Now you can see that ½ of the electricity escaped their notice.

"North and South Pole magnets are the Cosmic Force. The North and South Pole magnets are not only holding together the Earth and Moon, but they are turning the Earth around on its axis.

"Those magnets which are coming down from the Sun, they are hitting their own kind of magnets which are circulating around the Earth and they hit more on the East side than on the West side, and that is what makes the Earth turn around."

"Now about magnet size. Sunlight can go through glass, paper, and leaves, but it cannot go through wood, rock and iron, but the magnetism can go through everything."

"The results of North and South Pole magnet's functions I call magnetic currents and not electron currents or electricity because electricity is connected too much with those non-existing electrons."

Government officials, engineers, and scientists tried to get him to reveal his technology to no avail. He died in 1951, taking his secret with him. From the above writings it seems that the Cosmic Force and "Life Force" are one and the same. If he found the lost knowledge in an European library in recent times then the odds are that it still remains there to be re-discovered. Think of the implications of the use of this force. It would be a new clean power source to heat and power cities. It would be cheap and efficient. Nuclear and fossil fuel power plants would no longer be necessary. It might be used in aircraft as an anti gravitational device or power jet planes safer than the current use of flammable fuels, and could even be harvested in outer space to power spacecraft anywhere in the universe.

A new breed of automobiles, SUV's, and trucks could be designed to run on the new form of power. Builders could move heavy objects easily. Farmers could electrically charge their fields with the energy and produce bumper crops without the use of fertilizers. Most of all, the environmentalists would be in an extreme state of ecstasy rapidly approaching nirvana! These are just a few of the possible uses as the list goes on and on.

In John Michell's book, *The View Over Atlantis,* he states the following:

"We know that the whole surface of the earth is washed by a flow of energy known as the magnetic field. Like all other heavenly bodies, the earth is a great magnet, the strength and direction of its currents influenced by many factors including the proximity and relative positions of the other spheres in the solar system, chiefly the sun and moon. All the evidence from the remote past points to the inescapable conclusion that the earth's natural magnetism was not only known to men some thousands of years ago, but it provided them with a source of energy and inspiration to which their whole civilization was tuned."

In summary of this chapter it appears that the "Life Force" was known to the ancients by many names. According to the ancient writings it is a form of inter-dimensional energy that travels in spirals, whose power source comes from the giant rapidly spinning generator stars called Kolob and Oliblish. These generator stars are located near to the throne of God in another universe, called the "highest heaven", which is located in the highest spiritual universe above our temporal universe, in the fifth dimension. Since the great universes are so immense, it is logical to conclude that there may be other giant generator stars not mentioned in scriptures. The energy is filtered down through dimensions or universes and through various mediums this energy is " borrowed" by our sun and the other planets in our solar system. The energy permeates our temporal universe and affects all galaxies and all cosmic interlopers, such as comets and asteroids. It is the triggering force of creation that governs and keeps the great universes obedient to its laws. Modern day physics would equate the "Life Force" to an electromagnetic web of energy that interlocks all things both temporal and spiritual. Simply stated, the "Spiraling Life Force" is the pulse of the great universes.

Modern scientists have condensed the forces of the universe to a short list of four. Small and large molecular forces, electromagnetism, and gravity encompass all known forces in the universe. The Earth has a powerful magnetic field which diverts oceans of cosmic rays that bombard its atmosphere at speeds between four hundred and one thousand kilometers per second. Without that magnetic shield, life as we know it would cease to exist on the surface of the planet. The polarity of the magnetic flux moves millions of gallons of water through the skies above using an extremely powerful electrical movement called the electrojet. It is this field that Nikolai Tesla discovered and utilized in one fashion or another in hundreds of inventions. The fact that Earth is exactly the right distance from the sun, has a unique and effective shield against cosmic radiation, and revolves precisely enough to allow all life to propagate in all its forms is no mistake. The statement of mystery by many scientists as to the astonishing order with which the known universe operates is amazing. Even with all the answers provided by tools like the Hubble Space Telescope, the discovery of thousands of organized galaxies at the furthest and youngest reaches of its digital eye was not expected. "Why," they asked. "How could these galaxies be ordered and organized at such a *young* age?" The question should be, "Who is responsible for this order and organization?" Who indeed?

End Notes

[1] *(Teachings of The Prophet Joseph Smith, compiled by Joseph Fielding Smith, 1976, Deseret Book Company, Salt Lake City, pp.348-349)*

[2] (*1 Corinthians* 8:5-6)

[3] (*The Hebrew Book of the Dead,* translated by Zhenya Senyak, Tiktin Press, Hallandale, Fl. , 2003, p. 8)

[4] (Excerpted from *The Hebrew Book of the Dead In the Wilderness* ©2003 by Zhenya Sunyak. Published by Tikin Press, Hallandale, FLA. Used by permission of the publisher. All rights reserved. pp.131-132)

[5] Excerpts taken from *The Return of The Serpents of Wisdom* ©1997 by Mark Amaru Pinkham. Published by Adventures Unlimited Press, Kempton, ILL. All rights reserved. Used by permission of the publisher pp.275, 322,323,324)

[6] *Lost Cities & Ancient Mysteries of South America,* ©1986 by David Hatcher Childress. Published by Adventures Unlimited Press, Kempton, ILL. All rights reserved. , p. 25)

[7] (*The Return of the Serpents of Wisdom*, et al p. 322, 334, 335)

[8] (*The Return of the Serpents of Wisdom*, et al p. 337)

[9] *The Atlantis Blueprint,* @ 2000 by Rand Flem-Ath and Colin Wilson. Published by Delacorte Press a division of Random House, Inc. , NY, NY. p. 339)

[10] (*The Return of the Serpents of Wisdom*, et al p. 76, 78, 324, 334)

[11] (*The Return of the Serpents of Wisdom*, et al p. 333)

[12] (*The Return of the Serpents of Wisdom*, et al pp.190,191) (Jeremiah 38:1-5)

[13] (*The Return of the Serpents of Wisdom*, et al p323, 337)

[14] *Lost Cities & Ancient Mysteries of South America,* ©1986 by David Hatcher Childress. Published by Adventures Unlimited Press, Kempton, ILL. All rights reserved. p. 326)

The Garden of Eden

Almost all scriptures and ancient writings attest that the Garden of Eden was planted on the third day by God, eastward in Eden, on the spiritual earth, in the spiritual creation which is located in the fifth dimension.

To the north of the garden, there is a sea of water, clear and pure to the taste, like unto nothing else; the clearness thereof like unto crystal. One must enter the enclosed garden through the Gate of Paradise. The garden is described as beautifully planted with all manner of trees and flowers, that were pleasant to see and good for food. When the wind blew from the north, it carried the delicious smell of the trees of the garden over the spiritual earth. The fruits of the trees and vines were much larger than the fruits on the temporal earth; the weight of each fig being that of a watermelon. The leaves of the trees and plants never withered, bore blossoms and fruit continuously.

The Tree of Knowledge, located in the center of the garden, was like a species of the tamarind tree, bearing fruit which resembled grapes extremely fine; and its delightful fragrance extended to a considerable distance. The tree was enticingly beautiful in appearance and extremely pleasing to the eye. There was light, without any darkness, continually in the garden paradise. The Tree of Life, was also located in the center of the garden. The sweet odor it emits is beyond description and it is more beautiful than any tree growing in the garden. The tree is transparent as fire and glows like gold and crimson. Water issued forth from its root and formed a river that watered the garden. The river, after it left Eden, divided into four streams, which pour honey, milk, oil and wine, and then flowed out into the spiritual earth in four directions, winding about with a soft course. The name of the first river is Pison, the name of the second river is Gihon, the name of the third river is Hiddekel, and the name of the fourth river is Euphrates. In addition to the rivers, there was a mist that arose daily from the ground and watered the garden and the spiritual earth.

The above descriptions of the Garden of Eden are found in the Slavonic and Ethiopian texts of *The Book of Enoch*, and in the *Book of Adam and Eve*. An account of the four streams of Paradise, which pour honey, milk, oil, and wine, is also found in the *Qur'an* xivii.

Almost everyone is familiar with the Adam and Eve story. This is the most ancient story in the world. After Adam was created, God placed him in the garden to dress and keep it. God created spiritual animals and spiritual fowls of the air from the dust and waters of the spiritual earth and brought them to Adam to give them a name. An ancient belief states that all the animals, when first created, had the ability to speak, and all spoke one language. This belief still persists in modern times, with this additional "twist" each year on the day that the Christ child was born, at the stroke of midnight, the animals once again regain the ability to speak briefly. Maybe, the account in *Numbers* 22:28-30, of Balaam's she- ass, being smitten by her master, asked, "What have I done, unto thee, that thou hast smitten me these three times?" was perceived a normal event by Balaam because of the ancient belief among the Jews of his times that animals could possess the power of speech, under certain circumstances.

Later Eve was created to be Adam's wife, help meet, and companion. They were created in the beginning as immortal spiritual beings, in the image of God their maker. They were quickened or made alive by the Spirit of God, which flowed through their veins. Initially, Adam and Eve were created naked in the garden and were not ashamed but because their spiritual bodies were quickened by the Spirit of God, their appearance took on a "bright nature" which acted like a covering or robe of light to clad their spirits. These "robes or garments of light" are described many times in *The Zohar* as glorious vestures or raiment of glory that clothe righteous souls. There are different intensities or brightness of "light garments" for each level of glory or heaven attained by a soul. In the *Book of Adam and Eve*, Adam is described as a bright angel.[1]

The Lord God commanded Adam and Eve that they could eat of the fruits of every tree in the garden except of the tree of the knowledge of good and evil and gave them this warning; " for in the day that thou eatest thereof thou shalt surely die."

Now Lucifer, in the guise of a serpent, appeared to Eve in the garden and said; "Yea, hath God said, Ye shall not eat of every tree of the garden?" Eve replied, " We may eat of the fruit of the trees of the garden: But of the fruit of the tree which is in the midst of the garden, God hath said, Ye shall not eat of it, neither shall ye touch it, lest ye die." And the serpent said unto the woman, "Ye shall not surely die: For God doth know that in the day ye eat thereof, then your eyes shall be opened, and ye shall be as gods, knowing good and evil." And when Eve saw that the tree was good for food, and that it was pleasant to the eyes, and a tree to be desired to make one wise, she took of the fruit thereof, and did eat, and gave also unto her husband with her; and he did eat.

Many have incorrectly assumed that this was the first sin that brought death into the world. Adam and Eve were innocent when placed in the garden, and knew not right from wrong, so therefore were incapable of sinning but they were guilty of disobedience and aspiring to become as Gods. They had been warned by God and knew the consequences of eating the forbidden fruit. We can relate to this in today's times as often the most forbidden thing is the most desired and many will break the law in order to obtain the forbidden desired object, even at risk of going to jail, if caught. No where in Old Testament literature is found the doctrine of death entering the world through the sin of Adam and Eve. When they defied the Lord's command and ate of the forbidden fruit, their "garments of light" faded exposing their souls to no covering.[2] There was no way to conceal their actions; they were literally caught in the act. The *Genesis* account continues: "And the eyes of them both were opened, and they knew that they were naked; and they sewed fig leaves together, and made themselves aprons."

Later in the cool of the day, the Lord God, walking in the garden, called unto Adam, and said unto him, "Where art thou?"

Adam replied, " I heard thy voice in the garden, and I was afraid, because I was naked; and I hid myself."

The Lord again asked, "Who told thee that thou wast naked? Hast thou eaten of the tree, whereof I commanded thee that thou shouldest not eat?"

Adam answered, "The woman whom thou gavest to be with me, she gave me of the tree, and I did eat."

Then the Lord God cursed the ground with the following statement; "in sorrow shalt thou eat of it all the days of thy life; thorns also and thistles shall it bring forth to thee; and thou shalt eat the herb of the field; in the sweat of thy face shalt thou eat bread, till thou return unto the ground; for out of it wast thou taken: for dust thou art, and unto dust shalt thou return."

Other punishments were also meted out.

Then the Lord made coats of skins and clothed them. These coats of skins were coats of spiritual flesh, bone and blood, to house their souls, and they became mortal beings, subject to death; thus, the fleshless became spiritual flesh.[3]

To prevent Adam and Eve from eating of the Tree of Life, in the garden, and living forever, the Lord God sent them forth, from the Garden of Eden , to till the ground from whence he was taken. Cherubims, and a flaming sword, which turned every way to keep the way of the Tree of Life, prevented the couple from ever returning to the garden.

Adam and Eve were cast out of the garden into the lands of the spiritual earth. Because it was the spiritual earth, Adam and Eve were more physically like translated spiritual beings, not yet fully mortal in the temporal sense, but still subject to death. In *The Forgotten Books of Eden of Adam and Eve*, the couple commented on their physical appearance as follows: "But now we do not see as we did; our eyes have become of flesh; they cannot see in like manner as they saw before. What is our body today, compared to what it was in former days, when we dwelt in the garden?"

When they were expelled from the garden, Adam and Eve trod the ground on their bare feet not knowing that they were experiencing for the first time physical walking. The grace of a bright nature had departed from them and their hearts now turned toward earthly things.

The curse of mortality was also passed upon the animals, fowls, and plant life in the spiritual creation. Prior to Adam's transgression, there had been no death. The spiritual earth began to bare thorns and thistles to plague Adam and Eve in their efforts to till the ground by the sweat of their faces. The animals took on bodies of spiritual "flesh" and blood, similar in nature to the bodies of Adam and Eve. There was another, even more profound change. The bodies of Adam and Eve, once immortal and glorious, were also sterile. They had no children during all the time they were in the Garden of Eden. Was this one day, or was it one creative period of time? Some say a day to God is a thousand years on earth. If that's the case, then Adam and Eve were together for a thousand years during Day 7. We don't know when they partook of the Tree of Knowledge, but God and His Son did not return to visit them until the following day. So when Adam and Eve were remanded to mortality, their bodies were also enabled to bear children.

Adam and Eve began to have many children. Among their children were two sons, Cain and Abel. An account in *Genesis* 4 relates how Cain rose up and slew his brother, Abel, over jealously of acceptance of offerings, offered up to the Lord. This was the first committed murder and death on the spiritual earth. The Spirit of God continued to flow in the veins of Adam and Eve, and in their progeny, in combination with spiritual blood. When the Lord asked Cain, "Where is thy brother?" "What hast thou done?" The voice of thy brother's blood crieth unto me from the ground." It was the spiritual part of the blood that had the ability or voice to "cry out" to the Lord. It also was the main factor that enabled all born in the spiritual creation to live great spans of years. Adam lived nine hundred and thirty years, and as promised, by God, died in the day that he ate of the forbidden fruit; God's day being one thousand years.

Adam's son, Seth, lived nine hundred and twelve years. Enosh, son of Seth, lived nine hundred and five years. Cainan, son of Enosh, lived nine hundred and ten years. Mahalaleel, son of Cainan, lived eight hundred and ninety five years. Jared, son of Mahalaleel, lived nine hundred and sixty two years. Methuselah lived nine hundred and sixty nine years. Lamech, the father of Noah, lived seven hundred and seventy seven years.

We must take into account that a year on the spiritual earth may not be the same length as on the temporal earth; nevertheless, their lives were much longer than ours. It is doubtful if Adam fully realized what the consequences of death and returning to the earth's dust meant until Abel was slain. Adam was not illiterate. He could read and write. The following is found recorded in the book of *Moses* 6:5-6:

"And a book of remembrance was kept, in the language of Adam, for it was given unto as many as called upon God to write by the spirit of inspiration;

And by them their children were taught to read and write, having a language that was pure and undefiled."

Adam also kept a book of the generations or a genealogy of his children. *Moses* 6:8-9, as recorded states:

"And this was the book of the generations of Adam, saying: In the day that God created man, in the likeness of God made he him;

In the image of his own body, male and female, created he them, and blessed them, and called their name Adam, in the day

160

when they were created and became living souls in the land upon the footstool of God."

These books are also mentioned in *The Zohar*. [4]

The Zohar states "God sent a book down to Adam by the hand of the angel, Raziel, the angel in charge of the holy mysteries, from which he became acquainted with the supernal (supernatural, divine, or heavenly) wisdom. It came later into the hands of the "sons of God", the wise of their generation, and whoever was privileged to peruse it could learn from it supernal (supernatural, divine, or heavenly) wisdom. The book contained inscriptions revealing the sacred wisdom, and seventy-two branches of wisdom expounded so as to show the formation of six hundred and seventy inscriptions of higher mysteries. In the middle of the book was a secret writing explaining the thousand and five hundred keys which were not revealed even to the holy angels, and all of which were locked up in this book until it came into the hands of Adam. God sent the angel, Hadarniel, to warn Adam not to reveal the glory of the Master to the angels, for to him alone was given the privilege to know the glory of the Master. Therefore Adam kept the book by him secretly until he left the Garden of Eden. While he was there he studied it diligently, and utilized constantly the gift of his Master until he discovered sublime mysteries, which were not known even to the celestial ministers. This book was brought down to Adam by the "master of mysteries", preceded by three messengers. When Adam was expelled from the Garden of Eden, he tried to keep hold of this book, but it flew out of his hands. He there-upon supplicated

God with tears for its return, and it was given back to him, by the angel Raphael, in order that wisdom might not be forgotten of men, and that they might strive to obtain knowledge of their Master."

Adam left it to his son, Seth, who transmitted it in turn to his posterity, and so on until it was received by Enoch, the seventh from Adam. Tradition further tells us that Enoch also had another book, which came from the same place as the book of the generations of Adam. This is the source of the book known as *"The Book of Enoch"*. When God took him, He showed him all supernal mysteries, and the Tree of Life in the midst of the Garden and its leaves and branches, all of which can be found in his book. Happy are those of exalted piety to whom the supernal wisdom has been revealed, and from whom it will not be forgotten forever, as it says, "The secret of the Lord is with them that fear him, and his secret to make them know it."[5]

In the time of Enos/Enosh, son of Seth, men were skilled in magic and divination, and in the art of controlling the heavenly forces; the result of having the knowledge of *The Book of Divine Wisdom*. It is sometimes referred to as The *Book of Adam* and *The Book of the Angel Raziel*. He and his contemporaries studied the book and practiced magic and divination. From them these arts descended to the generation of the flood and were practiced for evil purposes by all the men of that time. The practice of these arts commenced with Enos/Enosh, and hence it is said of his time, THEN WAS THE NAME OF THE LORD CALLED UPON PROFANELY. [6] Enosh instituted the practice of chiseling images and idols with a "style" that corrupted the world.[7] In those days men served other gods. They made idols of brass, stone or wood, and bowed down to them. Because they forsook the Lord and did evil in his sight, the Lord caused the waters of the river Gihon to overwhelm and consume them. One third of the population on the spiritual earth was destroyed. In spite of God's judgment, the sons of men did not turn from idolatry or turn away from doing evil in his sight.[8] The death of Adam took place at the end of nine hundred and thirty years that he lived upon the earth; on the fifteenth day of Barmudeh, after the reckoning of an epact of the sun, at the ninth hour. It was on the sixth day (Friday), the very day on which he was created, and on which he rested; and the hour at which he died, was the same as that at which he came out of the garden.[9]

The Book of Jude reveals that Enoch was the seventh from Adam and was a great prophet of God. He foresaw not only the flood of Noah but also the second coming of the Messiah. (*Jude* 14,15) Enoch, the son of Jared, was a righteous man and beloved by God, who chose him above all men upon the earth to be his appointed scribe of His creation of visible and invisible things, to be an avenger of the sins of men, and a succor to his family. So beloved was Enoch, of God, that he was permitted to enter the Garden of Eden and see the Tree of Life. God sent angels to take Enoch to the different levels of the spiritual heavens or universes and to write a record of what he saw there. These writings contained many of the visions that he experienced of other heavens, the astronomy of the temporal earth, an allegorical forecast of the history of the world up to the kingdom of the Messiah, and a series of prophecies extending from Enoch's own time to about one thousand years beyond the present generation. Adams book of remembrance *(The Book of Adam)* and book of genealogy was passed down to Enoch for safe keeping.[10]

Enoch wrote down the descriptions of all the creations which the Lord had made. He wrote 366 books and gave them to his sons, with the instructions that they should give the books to others that they may know the works of the Lord, which are very wonderful; to know that the Lord is the creator of all; and to understand that there is no other God beside Him.[11]

Other books were given to Enoch from the Lord's store-places. An archangel, named Vretil, was appointed by the Lord to write down all the doings of the Lord. Vretil was commanded by the Lord to show and interpret to Enoch the books and let Enoch make copies. Enoch states that the wonderful books were fragrant with myrrh.[12]

God commanded Enoch to preach repentance and salvation to all the children of Adam. He was able to successfully combat the evils of the time and segregate and raise up a righteous culture called Zion, meaning, "the pure in heart." Enoch said that he came out from the land of Cainan, the land of my fathers, a land of righteousness unto this day and that his father taught him in all the ways of God. In time, Enoch built a city that was called the City of Holiness, even Zion. The people who lived there were of one heart and mind, dwelt in righteousness and there was no poor among them. This city was taken up into God's realm. For many generations there after, the Holy Ghost fell on the righteous Saints, and they were caught up by the powers of heaven into Zion, without experiencing death.[13]

God revealed to Enoch the age of the spiritual world, its existence of seven thousand years. Enoch was taken into the Lord's realm, without experiencing death in the month of Tsivan on the same day, the sixth day; the day on which he was born, and the same hour. He lived 365 years. Modern day revelation gives the 365 years as the age of Enoch when the city of Zion was taken into God's realm; and Enoch, without experiencing death, at the age of 430 years, was taken by God, into his bosom.[14] Before he was taken by the Lord, Enoch begat three sons; Methuselah, Elisha, and Elimelech (Eliakim); and two daughters; Melca and Nahmah (Naamah); and he was the great grandfather of Noah. (Chap. LXXXIV p. 121)

The Lord commanded Enoch's children to distribute the books to their children's children, and from generation to generation, and from nation to nation. Though the books were committed to the keeping of men, the Lord also instructed two special angels, Ariukh and Pariukh, to be the guardians of the writings so that they may not be lost in the deluge to come. They were further instructed to protect them until the last age and the time for their complete disclosure and understanding had come.[15]

In time, the children of Adam multiplied and spread over the spiritual earth. Cain, who slew his brother Abel, dwelt in the land of Nod and built the first city, on the spiritual earth, which he named Enoch.

Genesis 6:2, 4 records that, "the sons of God saw the daughters of men that they were fair; and they took them wives of all which they chose."

"There were giants in the earth in those days; and **also after that**, when the sons of God came in unto the daughters of men, and they bare children to them, the same became mighty men which were of old, men of renown."

These scriptures state that the giants were on the earth before and after the flood. *The Zohar* refers to them as Nephilim (lit. fallen ones) and from them the "mixed multitude" derive their souls, the result of falling into fornication with fair women. The "mixed multitude" consists of five categories: Nephilim, Gibborim, Anakim, Rephaim, and Amalekites. One group, called the Rephaim, were the terrible giant progeny resulting from the union of a class of angels called the watchers with the daughters of men. These watchers were from a class of angels known collectively as Nephilim. Not all watchers were Nephilim. The Nephilim were the angels cast out of heaven, along with Lucifer, when they warred against God. The Gibborim (mighty ones) are those of whom it is written: "they are the mighty ones...men of name" (*Gen.* 6:4). They come from the side of those who said "Come, let us build a city and make to us a name" (*Gen.* 11:4). These men erect synagogues and colleges, and place in them scrolls of the law with rich ornaments, but they do it not for the sake of God, but only to make themselves a name. They make up a part of the "mixed multitude."

When God thought of making man, the angels began to malign him and said, "What is man that Thou shouldst remember him, seeing that he will assuredly sin before Thee." God replied to them, "If ye were on earth like him, ye would sin worse." And so it was, for "when the sons of God saw the daughters of man", they fell in love with them, and God cast them down from heaven.[16]

In the *Ethiopian Book of Enoch The Prophet*, the sons of God (nephilim) are identified as fallen angels from the class of the watchers, who became enamoured with the beautiful daughters of men and decided to take them as wives and beget children. Their ring leader was Samyaza, along with 199 sons of God, devised, bound themselves by mutual execrations, and executed their projected undertaking. All 200 of them descended upon the top of mount Armon, on the spiritual earth. The names of their other chiefs, besides Samyaza, were Urakabarameel, Alibeel, Tamiel, Ramuel, Danel, Azkeel, Saraknyal, Asael, Armers, Batraal, Anane, Zavebe, Samsaveel, Ertael, Turel, Yomyael, and Arazyal. They took wives of their choosing and taught them sorcery, incantations, the workmanship of bracelets and ornaments, the use of paint to beautify the eyebrows and the use of stones of every valuable and select kind. Their women conceiving brought forth giants, whose stature was each three hundred cubits—12 to over 18 feet tall. These devoured all which the labor of men produced, until it became impossible to feed them. When they ran out of food, they began to devour men, birds, animals, reptiles, and fish. They ate their flesh and drank their blood. Azazyel taught men to make swords, knives, shields, breastplates and the fabrication of mirrors. Amazarak taught all the sorcerers, and dividers of roots; Armers taught the solution of sorcery; Barkayal taught the observers of the stars; Akibeel taught signs; Tamiel taught astronomy; and Asaradel taught the motion of the moon.

The whole earth was corrupted by the effects of the teaching of the work of Azazyel. He taught every species of iniquity upon the earth and disclosed to the world all the secret things which are done in the heavens. Men, being destroyed, cried out to God for deliverance. Their blood, which was shed on the spiritual earth, cried out to God for mercy and complained even to the gate of heaven.[17]

Genesis 6:5 states: "And God saw that the wickedness of man was great in the earth, and that every imagination of the thoughts of his heart was only evil continually."

The worst evil in the sight of the Lord was these two things: they sacrificed unto devils, not to God; to gods whom they knew not, to new gods that came newly up, whom your fathers feared not;[18] and the practice of a secret combination, in the dark, that was sodomy.[19]

The Ethiopian Book of Enoch continues: "The Lord heard the cries of tormented men and ascribed the whole crime to Azazyel. He sent his angel, Raphael, to bind Azazyel hand and foot and to cast him into an underground desert darkness. He is to remain there in darkness until the great day of judgment and then be cast into fire . To the angel Gabriel the Lord said, "Go to the biters, to the reprobates, to the children of fornication, the offspring of the

Watchers, from among men; bring them forth, and excite them one against another . Let them perish by mutual slaughter; for length of days shall not be theirs." To the angel Michael likewise the Lord said, "Go and announce his crime to Samyaza, and to the others who are with him, who have been associated with women, that they might be polluted with all their impurity. And when all their sons shall be slain, when they shall see the perdition of their beloved, bind them for seventy generations underneath this earth, even to the day of judgment Then shall they be taken away into the lowest depths of the fire in torments; and in confinement shall they be shut up for ever."

Before all these things happened, the Lord sent Enoch to Azazyel and to the other fallen Watchers of heaven, to deliver the following message: "That on the earth they shall never obtain peace and remission of sin. For they shall not rejoice in their offspring; they shall behold the slaughter of their beloved; shall lament for the destruction of their sons; and shall petition for ever; but shall not obtain mercy and peace."

The Watchers, hearing the message Enoch delivered, became terrified and beseeched Enoch to write a petition to the Lord in their behalf, that they might obtain forgiveness for all that they had done and obtain remission and rest. Enoch wrote up their petition but he was shown in a vision that judgment has been passed upon them. Their request would not be granted them as long as the world endures.[20]

God casts them out from the future world, in which they have no portion, and gives them their reward in this world—the original creation.[21]

Because of the great wickedness and iniquity upon the earth, the Lord revealed to Enoch, his plan to destroy the earth and every living creature upon it by means of a great flood; that a new world would be established and peopled by a new righteous race.[22] "For all flesh had corrupted their way upon the earth." This indicates that the whole of the animal world had become corrupted and had confounded their species. Observe that it was the wicked among mankind who brought about the unnatural intercourse in the animal world, and who sought thereby to undo the work of creation; they made the rest of creation pervert their ways in imitation of themselves. Said God to them: "You seek to undo the work of my hands; your wish shall be fully granted, for every living thing that I have made will I blot out from the face of the earth. I will reduce the world to water, to its primitive state, and then I will form other creatures more worthy to endure."[23]

In view of the above statement that the sons of God, their giant offspring, and the evil sons of men, were having sexual intercourse with the spiritual animals and confounding the species, suggests that perhaps the Greek legends of Minotaurs, Satyrs, Centaurs, and other part human-like, part animal beings, were indeed based on fact and not myths.

Before Enoch was taken by the Lord, he was given the promise by the Lord, that his posterity would remain forever on the earth, while the earth should stand, through Noah and his sons after the deluge.[24] He entrusted a copy of all the books he had written, and in addition, *The Book of Divine Wisdom, The Book of Adam's Remembrance* , and *The Book of the Generations of Adam*, to his son, Methuselah, for safe keeping.[25]

In time, Methuselah, son of Enoch, took a wife for his son, Lamech. Lamech was 180 years old when he married Ashmua, the daughter of Elisha, the son of Enoch, his uncle. She became pregnant by him, and brought forth a child, the flesh of which was as white as snow and red as a rose; the hair of whose head was white like wool, and long; and whose eyes were beautiful. When he opened them, he illuminated all the house, like the sun; the whole house abounded with light. When also he was taken from the hand of the midwife, he opened his mouth, and blessed the Lord of heaven.

Lamech, his father, was afraid of him and went straightway to his father and said, "I have begotten a son, unlike other children. He is not human; but, resembling the offspring of the angels of heaven, and is a different nature from us. And now, my father, let me entreat and request you to go to our progenitor Enoch, and to learn from him the truth; for his residence is with the angels."

When Methuselah heard the words of his son, he sought out Enoch at the extremities of the earth; for he had been informed that he was there: and he cried out.

Enoch heard his voice, and went to him saying, "Behold, I am here, my son; since thou art come to me."

Methuselah told Enoch of the strange child that had been born to his son, Lamech, and wanted to know if the child belonged to him or to one of the fallen angels. Enoch answered and said, "A great destruction therefore shall come upon all the earth; a deluge, a great destruction, shall take place in one year. This child which is born to you shall survive on the earth, and his three sons shall be saved with him. When all mankind who are on earth shall die, he shall be safe. Now therefore inform thy son Lamech, that he who is born is his child in truth; and he shall call his name Noah, for he shall be to you a survivor . He and his children shall be saved from the corruption, which shall take place in the world; from all the sin and from all the iniquity, which shall be consummated on earth in his days. Afterwards shall greater impiety take place than that which had been before consummated on the earth; for I am acquainted with holy mysteries, which the Lord himself has discovered and explained to me; and which I have read in the tablets of heaven. In them I saw it written, that generation after generation shall transgress, until a righteous race shall arise; until transgression and crime perish from off the earth; until all goodness come upon it."

When Methuselah heard the word of his father, Enoch, who had shown him every secret thing, he returned with understanding, and called the name of that child Noah; because he was to console the earth on account of all its destruction. Adam died when Lamech was 56 years old.[26]

In the *Messianic and Visionary Recitals of the Dead Sea Scrolls*, Noah was also "born perfect". Because of this "Perfection", Rabbinical literature has Noah born circumcised.[27]

Methuselah was a righteous man being of lineage of a righteous patriarchal line.[28] *Moses* 8:2 records that he was not taken with the city of Enoch so that the line could be continued and fulfill the covenant made with Enoch that Noah should be of the fruit of his loins. *Genesis* shows that he died in the year of the flood at the age of nine hundred and sixty nine years, however, once his work was done he may have been translated too, like the other righteous Saints in his time. *The Book of Jasher*, in chapter 5:21, states, "that all the good and faithful died 1 year before the flood, so as not to behold the evil that God would bring upon their brothers and relatives. That is why no believers were killed in the deluge." Noah was 195 years old when his father, Lamech, died; Lamech being 777 years old.[29]

Genesis 6:9 states that Noah was a just man and perfect in his generations, and Noah walked with God. He was a man endowed with wisdom and one who understands the secret mysteries of heaven as well as someone who knows the secrets of all living things. This knowledge and wisdom was passed down to him, by his great grandfather, grandfather, and father, and through the books that were passed down to his father, who in turn, passed them on to him for his safe keeping.[30]

As written in *The Book of Jasher*, the Lord said to Noah and Methuselah, "Stand forth and proclaim to the sons of men all the words that I spoke to you in those days, peradventure they may turn from their evil ways, and I will then repent of the evil and will not bring it. The Lord granted them a period of 120 years, saying, if they will return, then will God repent of the evil, so as not to destroy the earth. And Noah and Methuselah stood forth, and said in the ears of the sons of men, all that God had spoken concerning them, but the sons of men would not hearken, neither would they incline their ears to all their declarations."[31]

The sons of men proclaimed, "What is the Almighty that we should serve him?"[32] Relying upon the knowledge they had acquired from *The Book of Divine Wisdom*, in the arts of magic, divination, and in the art of controlling the heavenly forces, they defied Noah, saying that divine justice could never be executed upon them, since they knew a way to avert it. And furthermore, because they knew the angel in charge of fire and the angel in charge of water, they believed that they were safe because they had means and knowledge of preventing them from executing judgment on them. They only saw that the world was entrusted to those chieftains (sons of God) and that everything was done through them, and therefore they took no heed of God and His works, until the time came for the earth to be destroyed and when the Holy Spirit proclaimed every day, "Let sinners be consumed out of the earth and let the wicked be no more."[33]

Noah refrained from taking a wife and having children because of his knowledge of impending judgment from God, that would result in the earth's destruction, but the Lord commanded him to take a wife and beget children, in order that a righteous race could be preserved in the earth. When Noah was 498 years old, he took Naamah, the daughter of Enoch, who was 580 years old, to wife.[34] Modern day revelation reveals that Noah was 450 years old when he begat Japheth. Forty-two years later he begat Shem, and when he was 500 years old, he begat Ham.[35] *The Book of Jasher* agrees with *The Book of Moses* in the birth order of Noah's sons.[36] *Genesis* 5:32 gives the birth order as Shem, Ham, and Japheth. Interestingly enough *The Zohar 1*, page 246, agrees with *Genesis*. Does it really matter? Probably not, but it raises the question did one writing influence another? It is possible that the Hebrew *Bible* influenced *The Zohar* writings? It is doubtful if *The Book of Jasher* influenced *The Book of Moses*. *The Book of Moses* was revealed to Joseph Smith in June 1830, and it wasn't until 1840 that *The Book of Jasher* was translated from the original Hebrew into English. Obviously, *The Book of Jasher* had the ages of Noah and his wife incorrect, but nevertheless, they were of great age in comparison to ages of people in today's times. Again, a year on the spiritual earth may not correspond to a year of our time. It could be longer or shorter. The ancient records are not clear on this matter. Even today there are about forty different calendars used in the world, particularly for determining religious dates that differ from the Gregorian calendar most modern countries use.

The Christian calendar—Gregorian calendar—is based on the motion of the earth around the sun, while the months have no connection with the motion of the moon. On the other hand, the Islamic calendar is based on the motion of the moon, while the year has no connection with the motion of the earth around the sun. Finally, the Jewish calendar combines both, in that its years are linked to the motion of the earth around the sun, and its months are linked to the motion of the moon. The Chinese, Indian, and Japanese calendars are more complicated. The point is that time is measured according to where you live and it may not represent the same measurement used elsewhere.

Genesis 6:13 continues when God said to Noah, "The end of all flesh is come before me; for the earth is filled with violence through them; and, behold, I will destroy them with the earth.

Make thee an ark of gopher wood; rooms shalt thou make in the ark, and shalt pitch it within and without with pitch.

And this is the fashion which thou shalt make it of: The length of the ark shall be three hundred cubits, the breadth of it fifty cubits, and the height of it thirty cubits.

A window shalt thou make to the ark, and in a cubit shalt thou finish it above; and the door of the ark shalt thou set in the side thereof; with lower, second, and third stories shalt thou make it.

And, behold, I, even I, do bring a flood of waters upon the earth, to destroy all flesh, wherein is the breath of life, from under heaven; and every thing that is in the earth shall die."

The Ark of Noah was a barge-like structure built of gopher wood, a wood known on the spiritual creation but not known as such on the temporal creation. The word ark comes from the Hebrew word that means "box" or "chest". It was designed to float, not sail, and there were no launching problems. It may have had a bow and stern with a rudder. A cubit is a measurement of length, originally the length of the forearm from the elbow to the end of the middle finger or approximately 18 inches. The 18 inch cubit was known as the Hebrew cubit. An 18 inch cubit gives the measurements of the ark as 450 feet long, 76 feet wide and 45 feet tall, and are the figures most commonly quoted for its size. Here, the Hebrews have assumed that the Hebrew cubit was used in determining the dimensions of the ark. Did they forget that Noah wasn't a Jew and therefore would not have used the Hebrew cubit? Moses has been attributed as the inspired writer of *Genesis*. He was schooled in the Egyptian universities so therefore would have probably used the Egyptian cubit of 20.6 inches. The Hebrews also referred to a Divine Cubit of 25 inches. Was this cubit the one used in the spiritual creation? If so then the measurements would be 750 feet long, 125 feet wide, and 75 feet in height. There is an interesting "twist" to the dimensions of the ark. In the *Book of Adam and Eve*, page 68, it states:

"Now one spiritual cubit answers to the cubits of man, altogether forty-five cubits."

This was the spiritual creation and was the cubit measurement used there different from the cubit measurement used on the temporal creation? If the above spiritual measurement is so then the dimensions of the ark would be greater than the Hebrew, Egyptian, or Divine cubit. Were the Spiritual and Divine cubit one and the same unit of measurement?

The plot thickens when an ancient Babylonian historian and scholar, known as Berosus (280-261 B.C.), or referred to by Josephus as, "the Chaldaiaca", had access to ancient historical writings, no longer extant, and who said that the dimensions of the ark were five by two *stadia*, a common ancient unit of geographical measure of about 600 feet which is seldom used today. This would be 3,000 feet long and 1,200 feet wide. Berosus also gave a pre-flood list of kings that conforms to the traditional records of between eight and ten who reigned for between 186,000 and 456,000 years. He stated that Adam, the first of these kings, reigned for a period of 36,000 years. Scholars and theologians have perceived these numbers as errors in translations or else write off Berosus as "nuts". Berosus may not be as "nutty" as some have perceived him to be. Did he understand that Noah came from the spiritual creation, in another dimension, and that time and cubit measurements used there are reckoned by different celestial laws? Until it is revealed what measurement the Lord uses to make up a spiritual or celestial cubit, we can only speculate as to the dimensions of the ark. The ark was the largest ship built until the building of the iron clad ocean liners of today.

Noah was also commanded by God to choose wives for his sons, from the daughters of men. He did as God commanded and chose the three daughters of Eliakim, son of Methuselah, to be their wives.[37]

Genesis 6:18-22 continues with God stating to Noah:

"But with thee will I establish my covenant; and thou shalt come into the ark, thou, and thy sons, and thy wife, and thy sons wives with thee.

And of every living thing of all flesh, two of every sort shalt thou bring into the ark, to keep them alive with thee; they shall be male and female.

Of fowls after their kind, and of cattle after their kind, of every creeping thing of the earth after his kind, two of every sort shall come unto thee, to keep them alive.

And take thou unto thee of all food that is eaten, and thou shalt gather it to thee; and it shall be for food for thee, and for them."

Noah did as God commanded and commenced to build the ark in his five hundred and ninety-fifth year of age. It took him five years to build the ark.[38] How did Noah and his sons build the ark in such a short span of time? Some have speculated that Noah hired some locals to help and other writings indicate that he had angelic assistance. Once again, five years of spiritual time is not the same as five years here. It was probably longer. Another source indicates that the year was shorter. It could still have consisted of a number of months, but not of months of thirty days; and the days themselves could have been shorter. It all boils down to the fact that we do not know at present writings how time was measured in the spiritual creation. The months could have been more and shorter or just the reverse; fewer months and longer and we do not know how many hours constituted a day. In view of this, there is no choice but to accept a year there as being relative and not necessarily equal to our time.

One can imagine the mockery inflicted on "crazy" Noah and his family for building a ship in the middle of nowhere; not even near water. There had not even been rain upon the planet, but rather heavy dews that kept everything green in a climate of near equilibrium. Perhaps there were even death threats from preaching repentance for 120 years such as, "Kill him and let us be rid of his constant preaching." Jeers, name-calling, and shouts of "liar" were endured.

The *Book of Jasher* gives more insight than *Genesis* in the following writings:

"And it was at that time Methuselah, the son of Enoch, died, 960 years old was he, at his death. At that time, after the death of Methuselah, the Lord said to Noah, "Go thou with thy household into the ark; behold I will gather to thee all the animals of the earth, the beasts of the field and the fowls of the air, and they shall come and surround the ark. And thou shalt go and seat thyself by the door of the ark, and all the beasts, the animals, and the fowls shall assemble and place themselves before thee, and such of them as shall come and crouch before thee, shalt thou take and deliver into the hands of thy sons, who shall bring them to the ark, and all that will stand before thee thou shalt leave."

"And a lioness came, with her two whelps, male and female, and the three crouched before Noah, and the two whelps rose up against the lioness and smote her, and made her flee from her place, and she went away, and they returned to their places, and crouched upon the earth before Noah. And Noah saw this, and wondered greatly, and he rose and took the two whelps, and brought them into the ark. And Noah went and seated himself by the door of the ark, and all flesh that crouched before him, he brought into the ark, and all that stood before him he left on the earth. And the lioness ran away and stood in the place of the lions.

Two and two came to Noah into the ark, but from the clean animals and clean fowls, he brought seven couples, as God commanded him. And Noah brought into the ark from all living creatures that were upon earth, so there was none left but which Noah brought into the ark. And all the animals, beasts, and fowls that were remaining, they surrounded the area around the ark.

And on that day, the Lord caused the whole earth to shake, and the sun darkened, and the foundations of the world raged, and the whole earth moved violently, and the lightening flashed, and the thunder roared, and all the fountains in the earth were broken up, such as was not known to the inhabitants before; and God did this mighty act, in order to terrify the sons of men, that there might be no more evil upon earth. And still the sons of men would not return from their evil ways, and they increased their anger of the Lord at that time, and did not even direct their heart to all this."[39]

The ark was loaded with its precious cargo. The first level of the ark was probably food and provisions for all the animals. All the animals, and beasts prior to the deluge were herbivorous, and all the righteous people were vegetarians. The giant off spring of the sons of God however, did kill animals and men and ate their flesh. The second level was probably where the animals and beasts were kept in stalls. The top level was probably living space for Noah and his family. They would have kept food and household provisions for themselves on this level to last for a year. Noah also took aboard all the sacred books entrusted into his care. These books were not like books we have today, but were sacred writings wound on a scroll, called sticks.

Jared, when he was about to die, called Enoch his eldest son, and Methuselah, Enoch's son, and Lamech, the son of Methuselah, and Noah the son of Lamech, and made the following last directive to be carried out:

"And unto him of you who shall be left, O my son, shall the Word of God come...." Then Noah said unto him, "Who is he of us that shall be left?" And Jared answered, "Thou art he that shall be left. And thou shalt take the body of our father Adam from the cave, and place it with thee in the ark when the flood comes. And thy son Shem, who shall come out of thy loins, he it is who shall lay the body of our father Adam in the middle of the earth, in the place whence salvation shall come." Jared instructed them to take three precious gifts and offerings, namely, the gold, the incense, and the myrrh, and let them be in the place where the body of our father Adam shall lay.[40]

According to *The Zohar* Shem did take aboard the ark the bones of Adam and Eve.

Besides the books and Adam and Eve's bones, there was one more significant item taken aboard the ark. It was a sapphire stick called the Rod of God because the name of the Lord God of hosts was engraved upon it. This is the stick with which all the works of our God were performed, after he had created heaven and earth, and all the host of them, seas, rivers and all their fishes.

When Adam was driven from the Garden of Eden, he took this stick or staff with him and later it was passed down to Noah. [41] God commands Noah, his sons, and his wife, and his sons' wives with him, into the ark, because of the waters of the flood: and the Lord shut him in. (*Genesis* 7:7,16) Noah didn't enter the ark until after the Lord had covenanted with him. This makes the ark, symbolically, another ark of the covenant.

The faith of Islam holds that the Hebrew text has been corrupted through many translations and that the *Qur'an* came forth to restore the truth. They view that Noah's wife was not a believer with him, so she did not join him; neither did one of Noah's son, who was secretly a disbeliever but had pretended faith in front of Noah. Likewise most of the people were disbelievers and did not go on board. However, they believe that there were a small number of other believers aboard that Noah had converted.

This in the *Qur'an* suggests that there were indeed a few others, besides the family of Noah who believed him:

"The ship was constructed, and Noah sat waiting Allah's command. Allah revealed to him that when water miraculously gushed forth from the oven at Noah's house, that would be the sign of the start of the flood, the sign for Noah to act.

The terrible day arrived when the oven at Noah's house overflowed. Noah hurried to open the ark and summon the believers. He also took with him a pair, male and female, of every type of animal, bird and insect. Seeing him taking these creatures to the ark, the people laughed loudly: Noah must have gone out of his head! What is he going to do with the animals?

Almighty Allah narrated: (So it was)

Till our decision was implemented and water welled out from the oven. We said to Noah: Take into the ark a pair from every species, your family, except those against whom a word has already been passed and those who believe. But none except a very few believed him."[42]

The scholars (Ulama) hold different opinions on the number of those who were with Noah on the ship. Ibn Abbas stated that there were eighty believers, while Ka'ab Al-Ahbar held that there were seventy-two. Others claimed that there were ten believers with Noah.

Allah told the story in the *Qur'an* as follows : And he (Noah) said: "Embark therein, in the Name of Allah will be its moving course and its resting anchorage. Surely, my Lord is oft-forgiving, most merciful. So it (the ship) sailed with them amidst the waves like mountains, and Noah called out to his son, who had separated himself (apart), O my son! Embark with us and be not with the disbelievers. The son replied: I will betake myself to a mountain, it will save me from the water. Noah said: This day there is no savior from the decree of Allah except him on whom He has mercy. And a wave came in between them, so he (the son) was among the drowned.

And Noah called upon his Lord and said, O my Lord! Verily, my son is of my family! And certainly, your promise is true, and you are the most just of the judges. He said: O Noah! Surely, he is not of your family; verily, his work is unrighteous, so ask not of Me that of which you have no knowledge! I admonish you, lest you be one of the ignorants.

Noah said: "O my Lord! I seek refuge with you from asking you that of which I have no knowledge. And unless you forgive me and have mercy on me, I would indeed be one of the losers."[43]

Arab tradition holds that Yam (Kana-can) was the name of the son of Noah who perished in the deluge along with his mother, Waliya, who called Noah "madjnun" (insane). Was Waliya, an unbeliever, another wife of Noah or is Waliya an Arab name for Naamah?

All other sacred writings state that there were only eight righteous souls aboard the ark; Noah, his wife, his three sons and their wives. It is interesting to note that nothing more is said of Noah's wife, after the ark landed and they disembarked.

There are several other texts, not considered sacred, that name others aboard the ark, that predate the writings of Moses in *Genesis*. The Babylonians refer to two sailors, Buzur and Uragal, who both helped in handling the anchor lines for the ark's "braking stones." In the *Epic of Gilgamish*, there was a boatman, who may be Urshanabi, that brought aboard the shars of oil and who stayed with Utnapishtim (Noah) in his old age. Another version of the text, the *Sin-Lequi-Unninni*, says that even "the children of all the craftsmen I drove aboard." The tomb of Noah's sister is located in Syria and the tomb of Noah's mother is located in Iran. This would indicate that Noah had some of his close kin on board. The Babylonian records also indicate that Noah was to embark with his kin, closest friends, men servants, maid servants, and their young.

It would be logical for a vessel, the size of an ocean liner, to be manned by a crew of experienced boatman, who would also share in the daily care of thousands of animals. For a crew of only 8 souls to perform all of the necessary daily tasks would be practically impossible. Some theologians have suggested that the animals were in a state of hibernation and didn't require much care. If this were the case, then why did the Lord command Noah to take on board food for the animals? The whole lower deck was for animal food and plenty of it!

The Zohar writings state the following: "Meanwhile Noah was hidden in the ark, concealed from sight, so that the destroyer could not come near him or his family. You may perhaps say that there was not here any destroying angel, but only the onrush of the overwhelming waters. This is not so; no doom is ever executed on the world, whether of annihilation or any other chastisement, but the destroying angel is in the midst of the visitation. So here there was indeed a flood, but this was only an embodiment of the destroyer who assumed its name. Hence the command given to Noah to hide himself and not to show himself abroad. But, you may object further, the ark was exposed to full view in the midst of the world through which the destroyer was roaming. The fact is that this made no difference, since as long as the face of a man is not seen by the destroyer, he has no power over him. We learn this from the precept given at the time of the exodus, "and none of you shall go out of the door of his house until the morning."[44] The reason is that the destroyer was then abroad with power to destroy anyone who showed himself. For the same reason Noah withdrew himself and all under his charge into the ark, so that the destroyer had no power over him. This concealment fulfills the verse, "Seek righteousness, seek humility, it may be ye shall be hid in the day of the Lord's anger."[45] Before he entered the ark Noah caught sight of the angel of death going among the people and encircling them. As soon as he espied him, he went into the ark and hid himself, thus this verse: A prudent man seeth evil and hideth himself."[46]

Genesis 7:10-11 continues:

"And it came to pass after seven days, that the waters of the flood were upon the earth. In the six hundredth year of Noah's life, in the second month, the seventeenth day of the month, the same day were all the fountains of the great deep broken up, and the windows of heaven were opened."

The *Book of Jasher* adds to the account as follows: "And all the sons of men that were left upon the earth, they became exhausted through evil on account of the rain, for the waters were coming more violently upon the earth, and the animals and beasts were still surrounding the ark. And the sons of men assembled together, about seven hundred thousand men and women, and they came unto Noah to the ark. And they called to Noah, saying, open for us that we may come to thee in the ark, wherefore shall we die? And Noah, with a loud voice, answered them from the ark, saying, have you not all rebelled against the Lord, and said that he does not exist? And therefore the Lord brought upon you this evil, to destroy and cut you off from the face of the earth. Is not this the thing that I spoke to you of one hundred and twenty years back, and you would not hearken to the voice of the Lord, and now do you desire to live upon the earth?

And they said to Noah, we are ready to return to the Lord; only open for us that we may live and not die. And Noah answered them, saying, behold now that you see the trouble of your souls, you wish to return to the Lord; why did you not return during these hundred and twenty years which the Lord granted you as the determined period? But now you come and tell me this on account of the troubles of your souls, now also the Lord will not listen to you, neither will he give ear to you on this day, so that you will not succeed in your wishes.

And the sons of men approached in order to break into the ark, to come in on account of the rain, for they could not bear the rain upon them. (note: prior to this time, there were no seasons, cold, heat, snow or rain on the spiritual earth; only one mention of the cool of the day in the Garden of Eden)

And the Lord sent all the beasts and animals that stood around the ark. And the beasts overpowered them and drove them from that place, and every man went his way and they again scattered themselves upon the face of the earth.

And the rain was still descending upon the earth, and it descended forty days and nights, and the waters prevailed greatly upon the earth; and all flesh that was upon or in the waters died, whether men, animals, beasts, creeping things or birds of the air; there only remained Noah and those that were with him in the ark. And the waters prevailed and they greatly increased upon the earth, and they lifted up the ark and it was raised from the earth."[47]

From *The Zohar* we learn that rain descended upon them from above and at the same time scalding water, hot as fire, gushed up from below. The manner of their death was as follows: scalding water spurted up from the abyss, and as it reached them it first burnt the skin from the flesh, and then the flesh from the bones; the bones then came asunder, no two remaining together, and thus they were completely blotted out from the earth.[48]

Here is a description of the flood from the Islamic point of view: "Water rose from the cracks in the earth; there was not a crack from which water did not rise. Rain poured from the sky in quantities never seen before on the earth. Water continued pouring from the sky and rising from the cracks; hour after hour the level rose. The seas and waves invaded the land. The interior of the earth moved in a strange way, and the ocean floors lifted suddenly, flooding the dry land. The earth, for the first time, was submerged."

By the time Adam died there were seven generations of his own off spring on the earth. By a conservative count, the population was 120,000 persons minimum. By the time of the flood it is estimated that there were possibly as many as seven billion people, or higher.

Nowhere in the preflood accounts is any mention made of race or multiple languages. Presumably they were all of one race and spoke one tongue.

The time of the deluge is difficult to determine due the fact that the flood occurred on the spiritual earth, in the fifth dimension. More will be said on this topic in the chapter, *The Arrival of Noah*.

So what happened to the Garden of Eden? It is believed to have been taken into God's realm before the deluge and will remain there until the world to come, at which time it will return to the future world for all the righteous to enjoy.

Noah had embarked on a voyage of a lifetime. He literally sailed from the old world into a new world. For now we close on this chapter, but it will pick up where we left off under the chapter titled, *The Arrival of Noah*. In order to understand how this happened, one must read the next chapter as a prelude or else become completely lost.

In presenting this chapter most of the original text of the scriptures and sacred writings was used whenever possible, with few word changes and some spelling corrections. The phrasing was written in the "style" of that period of time. To alter it would take away the intent of the writer of that particular text and do away with some of the drama and beauty of the language of that particular age and time.

The flood did happen. It occurred on the spiritual earth, the original creation, in the fifth dimension. Peter, the Apostle, speaks of strong skeptics, "mockers" or "scoffers", and warns that these secular and religious "authorities" would appear on the scene at the end of the age we now live in and will vigorously deny that the flood of Noah ever happened, and that it is just a myth. Does this sound familiar?

End Notes

[1] *(Excerpted from The Forgotten Books of Eden ©1927 by Alpha House, Inc. Published by World Publishing, Nashville, TN. Used by permission of the publisher. All rights reserved. , chapter X, p. 10)*

[2] (*The Zohar,* translated by Simon, Sperling and Levertoff, ©1984, London, The Soncino Press, Ltd., Vol. 1 p. 136)

[3] (*The Zohar* et al Vol. V pp.346-347, Job 10:11, The Zohar V1, et al p. 136, The Zohar et al IV, p. 281)

[4] (et al Vol. 111, page 219)

[5] (*The Zohar* et al 1, p. 139)(The Zohar 1, et al pp.176,177)

[6] (*Zohar* et al 1, 179)

[7] (*The Zohar* et al IV, p. 149)

[8] (Excerpts taken from *The Book of Jasher* (1840), ©1988 by Artisan Sales, Artisan Publishers, Muskogee, OK. Used by permission of the publisher. All rights reserved , chap. 2:3-6)

[9] Excerpted from *The Forgotten Books of Eden* ©1927 by Alpha House, Inc. Published by World Publishing, Nashville, TN. Used by permission of the publisher. All rights reserved. Chapter IX, p 67.

[10] Excerpts taken from *The Book of Enoch The Prophet,* ©2000. Translated by Richard Laurence. Published by Adventures Unlimited Press, Kempton, ILL. All rights reserved. Used by permission of the publisher. p. xxiii)(*Moses,* 6:46)(*The Zohar* et al 1V p. 345-346)

[11] Excerpts taken from *The Book of The Secrets of Enoch,* ©1999 by The Book Tree. Published by The Book Tree, Escondido, CA. All rights reserved. Used by permission of the publisher. pp.29, 83)

[12] (The Secrets of Enoch et al pp.28, 29)

[13] (KJV *Moses 7*)

[14] (*Moses 7:69, Moses* 8:1) (*Hebrews* 11:5)

[15] (*The Book of the Secrets of Enoch,* et al pp.29,48,49,62,79,83,84)

[16] (*The Zohar,* translated by Simon, Sperling and Levertoff, ©1984, London, The Soncino Press, Ltd., 1, pp.99-100)

[17] (*The Ethiopian Book of Enoch the Prophet,* pp.5-10)

[18] (*Deuteronomy* 32:17, *The Ethiopian Book of Eno*ch, chap. xix p. 26)

[19] (*Moses* 5:51, *The Slavonic book of The* Secrets of Enoch, Chap. xxxiv p. 49)

[20] (*Ethiopian Book of Enoch* pp.6,7,8,9,10,11,12,14,15) (11Peter Chap. 2:4, The Zohar et al V p. 312)(Jude 6:6)

[21] (*The Zohar* et al 1 p. 99)

[22] *(The Secrets of Enoch,* et al chap. xxxiv 9. 49,50. Appendix p. 87,92)

[23] (*The Zohar* et al 1 p. 225-226)(Excerpts taken from *The Book of Jasher* (1840), ©1988 by Artisan Sales, Artisan Publishers, Muskogee, OK. Used by permission of the publisher. All rights reserved chap. 4:18)

[24] Excerpts taken from *The Book of The Secrets of Enoch,* ©1999 by The Book Tree. Published by The Book Tree, Escondido, CA. All rights reserved. Used by permission of the publisher. Appendix, p. 87, Moses 7:51,52)

[25] (*The Ethiopian Book of Enoch the Prophet,* chap. LXXX1, p. 114)

[26] (Excerpts taken from *The Book of Enoch The Prophet,* ©2000. Translated by Richard Laurence. Published by Adventures Unlimited Press, Kempton, ILL. All rights reserved. Used by permission of the publisher. ,pp174,175,176,177,178) (The *Book of Jasher* et al chap. 3:14, Chap. 4:11)

[27] (*The Birth of Noah* (4Q534-536) p. 33)

[28] (et al *Moses* 6:23)

[29] (Excerpts taken from *The Book of Jasher* (1840), ©1988 by Artisan Sales, Artisan Publishers, Muskogee, OK. Used by permission of the publisher. All rights reserved chap. 5:19, 20, 21)

[30] (*The Dead Sea Scrolls Uncovered, Messianic and Visionary Recitals, The Birth of Noah* (4Q534-536), p. 34) (*The Zohar 1* p. 177)

[31] (The *Book of Jasher*, et al chap. 5:8,11, 22-25)

[32] (*Job* XX1, 15)

[33] (*Ps.* CIV, 35) (*The Zohar* 1, pp.179,180)

[34] (The *Book of Jasher* et al chap. 5:14-18)

[35] (et al Moses 8:12)

[36] (The *Book of Jasher* et al chap. 5: 16-17)

[37] (The *Book of Jasher* et al chap 5:32-35)

[38] (The *Book of Jasher* et al chap. 5:33-34)

[39] (The *Book of Jasher* et al Chap. 5:36, Chap. 6:1-13)

[40] Excerpted from *The Forgotten Books of Eden* ©1927 by Alpha House, Inc. Published by World Publishing, Nashville, TN. Used by permission of the publisher. All rights reserved. Chap. XXI, pp.78-79)

[41] (The *Book of Jasher* et al p. 221)

[42] (et al Surah 11:40)

[43] (Surah 11:45-47)

[44] (Ex. X11:22)

[45] (*Zeph.* 11:3)

[46] (*Prov.* XX11:3) (*The Zohar* 1, p. 223,205,228,231)

[47] (*The Book of Jasher*, et al chap. 6:16-27)

[48] (*The Zohar* et al 1, pp.218,228)

The Cataclysm

The peaceful bejeweled expanse of space has always fascinated man. Since the first sentient eyes peered up into the night, beyond the glow of firelight, the patterns of stars and galaxies visible to the naked eye have given birth to myths and gods and somehow connected the mind behind those eyes to the past. Only now do we even speculate on how distant that past might be. The light reflecting off the most powerful mirrors into modern image processors may be the story of sources billions of years old. Still, astronomers stare into oculars and review still-shots of sectors in space hoping, hoping beyond hope to find something out of place— something different enough so as to indicate movement.

We have inherited the witness of every circa's meticulous observer. The sun and nearly all the planets were carefully plotted and recorded by every dynasty, cave community, and scribe. Some observers recorded only three planets and a sun. Some recorded four. Eleven planets, ten planets, and now only nine planets recorded for all time in these many records. So it is now. So it was then. Or, is the present a deceiving indicator of the past. One rule is the same for all times. One minute things are still and predictable, and the next minute chaos.

A planet revolves on its axis, and in its course is observed to rotate around another body in space, usually large enough to hold some governing force over the circumnavigating traveler. But, this is not always the case. Planets can have long elliptical orbits. In our solar system all the orbits of the planets lie practically in a single plane. Of course, this means if any one planet had anything but a relatively circular orbit the result might be conflict on a global scale. We now have evidence of tens of millions of planets that float through space without any orbit at all. These free-floating planets inevitably collide with something. Sometimes that collision is no more eventful than a fly striking the windshield of a jumbo jet. Recently we witnessed a shattered comet slamming into Jupiter at hundreds of thousands miles an hour generating hundreds of megatons in explosive power. The hole in Jupiter's atmosphere was visible for months, but total event left hardly any impression on the planet due to its immense size. Even further evidence is emerging that planets of like size have collided, or nearly collided.

Light speed has been theorized to be constant and the limiting velocity in the universe. Although this has been disproved and contradicted hundreds of times with subatomic particles accelerated to many times that speed, we shall capitulate and agree, for now. Let's suppose a jet aircraft is low to the ground and moving at nearly the speed of sound from behind you. You would be completely unaware of the approaching craft until it was well past vertical and visible down the exhaust of its engine. One second it wasn't there, the next, you've spilled your drink and ducked your head. So it is with planets moving through space, it seems. Earth moves at a leisurely 18.5 miles per second around the sun, but free-floating planets may be sling-shot many thousands of times that speed across the galaxy.

Histories from every surviving observer the world over recorded in their own languages and with their own frames of reference an event that is irrefutably real and that nearly killed all life on the planet earth. It first appeared as a distant glow. The first observers noticed it was not the normal whitish or bluish sparkle of a familiar planet or a star. This was a red glow like the color of blood. Barely visible to the naked eye, its careening path—hardly a revolution like a whole planet—appeared to form a tail not unlike that of a comet as it approached the sun and boiled off some of the frozen sea chunks pulled along by its glowing head. Over a period of weeks it drew the attention of every continent in its line of sight. Of course it made no sound in the void of space. It seemed to stay in the same place in the night sky, but it continued to grow in size and color. That's because it was on a collision course with the point of observation; the temporal earth.

The village's best night eyes must have climbed the nearest mountains to get the best view possible. Through their fear and surprise they must have reported the terrible form and action of the body hurling through space toward their retinas. Within days the great and silently terrible event was visible in the noon-day sky. According to dozens of records kept by historians all over the world the approaching body was not alone, nor was it an unfamiliar shape. As it rotated in space, ever changing arrangement and perspective like picturesque clouds on a lazy afternoon breeze, it galloped forward like a red dragon or a serpent with a fifty-thousand mile tail. First its fierce mouth would gape open wide after its prey. Then its menacing claws would lash outward, flickering with bright molten orange talons. Some recorded that as earth's perspective changed the crescent-shaped coma around the blazing head looked like a giant bull with red horns. The Chaldeans described it as a "bright torch of heaven."[1] The Mexicans called it a "smoking star," or Tzonte-mocque. The Aztecs called it Quetzal-cohuatl, commonly known as the feathered serpent. The Egyptians called it Zebbaj or "one with hair," as did the Babylonians.[2] Some scientists say this was the precursor for Venus. One translation of the word Venus is *hairy planet*. Some say it was Marduk, the victorious planet that slammed into Tiamet and broke it into the large pieces. As it traveled through our solar system it's gravity drew shattered remnants that trailed behind and eventually fell into our oceans and across our continents. But, in the beginning ages of human record, authors writing on clay or gold or paper believed in heavenly purposes to the motions in space. The panic and chaos on earth was only beginning.

The evidence of at least a near pass from another planet or very large comet is clear. Physicists have calculated the mechanics of this planet-to-planet interaction. The Roche limit is the orbital distance at which a satellite with no tensile strength (a "liquid" satellite) will begin to be tidally torn apart by the body it is orbiting. [3] In other words, as two bodies in space approach one another the gravitational forces that hold the smaller or less dense body together begin to be overcome by the gravitational field of the more dense planet. Keep in mind that planets do not have a perfectly spherical gravitational field. Rotating planets have centrifugal gravity. They also have internal cores that rotate at different speeds than the crust. The result is a term called *effective gravity*. This often takes the shape of large lobes of gravity that protrude millions of miles into space. Whenever two planets are near one another, these lobes, termed *Roche Lobes,* may intersect at points known as *Lagrange Points*, a fairly stable orbit may result. But, when the Roche Limit is encroached the result is spectacular. Even the larger planet undergoes massive changes to its rotational speed, and crustal features such as mountains and oceans.

As the remnants of what could have been a shattered planet came close and applied its gravity, the earth's crust heaved and moaned in agony. Volcanoes erupted. Once peaceful and majestic mountain ranges split open and bled destroying lava from their souls to the sea. Valleys vomited shards of granite and basalt high in to the sky at all angles. Soon the seas, once governed with the regularity of eternity by the moon, began to reach out to the oncoming invader. Within days the crust of the earth began to slow its daily rotation around the earth's core. When the crust stopped, half the earth was in darkness that lasted for at least 3 days; freezing and rumbling. The half facing the sun or the approaching body burned to the dirt. Like a cosmic generator the solid momentous core of the earth kept spinning, launching the magnetosphere a hundred thousand miles further into space. Its attraction to the twisted careening core of the approaching maestro was focused twenty-three and one-half degrees askew of earth's north pole. The magnetic flux lines pouring out of the top of the world aligned with the elliptical path of the passing serpent-bull. The planet tilted in realignment. The new north—the current 23 degrees off-center—would change the face and the life of the planet forever. Rocks formed along the fjords of Scandinavian countries from molten lava before and after this event were frozen with different magnetic norths, indicating that Earth had one direction for north before and a new direction for north after the cataclysm.

The oceans stood thousands of feet into the sky—spilling as far as the laws of physics would allow toward their new and powerful attraction leaving the rest of the world trembling and violated. Earth's poles shifted and waned for days while the side of the earth toward the sun raged with fire and the leeward side blazed with ice storms. Flora and fauna perished by entire species. The life of the seas once free to roam the globe now thrashed in geotropic confusion. Muck made of slurried animals, plants, and men heretofore not thought to coexist on Earth was jammed into cliff faces, crevices, caves, and fissures all over the planet. There bones and fossilized cellulose has been exhumed and carefully dated. Even the omnivorous man was nearly swept from the face of the earth as the now tilted and wobbling earth began to restart its rotation, due to the momentum of the dense planet core dragging the core into motion, and the Roche intruder pulling at the tangent of its exit from the earth's surface.

As the planetary puppeteer moved closer, the electrical potential between the planets was unleashed. Dozens of civilizations recorded these events in their writings, paintings, and legends. Gigantic bolts of lightning danced like flickering swords splitting space and sky. Some looked like spears or swords. Some looked like whips or arrows. The length of each blast was immense. The thunder rumbled the ground so violently that everything fell down. Trees at ground zero were either snapped over like wheat before the storm, or they were vaporized in a flash of infrared fury. Each pulse of the electricity surging into the earth turned thousands of tons of rocks and dirt into pure energy. Mountains of debris were launched into the stratosphere, catching the chaotic jetstream on a global path to blot out the sun. The resulting shade would last for 50 years or more. Blast after blast, each discharge took on a new sound. Witnesses record some sounding like long and terrible blasts from a cosmic trumpet. Others said they heard a voice—deep and foreboding—naming the deliverer of the cataclysm.

Within a few days the serpent's head would continue on its path, narrowly avoiding the limit of its own destruction. But, the worst was yet to come. Its entourage of smaller debris, doubtless the frozen remnants of the shattered planet's sea and atmosphere, was caught up in earth's gravitational pull. Striking the atmosphere at thousands of miles an hour, the smaller pieces exploded. Rains on a global scale fell in rivers onto the barren and scorched land. The larger pieces struck the earth hard enough to split the crust. The ocean floor ripped open half way across the planet. Hundreds of smaller pieces, laden with billions of tons of iron, splattered across the equator, the Carolinas, the Atlantic, and across Europe into southern Siberia. The ocean floor was covered in three feet of iron pellets, frozen as they struck the ocean like white-hot hailstones. Forests were flattened, and lakes were created with each impact. At the same time, the deeper oceans began to recede from their dance in the sky. Tides as high as six thousand feet traveled at speeds over one hundred miles an hour, crashed around the planet, and slammed into one another on the other side. It took months for the level to equilibrate. The polar caps formed during decade-long ice storms under the blotted solar energy.

The celestial visitor lumbered toward the edge of the solar system, prophesied to return again someday. A new planet appeared in the eastern sky. A morning star, now with the most circular orbit of any planet in the solar system, Venus stands bright and regular and peaceful to behold. But, the earth would never be the same. The once peaceful and ever tropical planet was forever to be turbulent and extreme with its new-found attitude toward the sun. The moon was pulled thousands of miles further into space. The rotation of the earth, now slower, wobbled with an instability that would curse the crust with ever shifting tectonic chaos. The oceans, now mixed with the sea life of a distant planet, battled for millennia to a new ecology. The atmosphere, once thick with ozone and billions of gallons of protective moisture, now condensed with a tiny dew point, would expose the planet to deadly cosmic radiation 50 years after the dirty sky was cleansed with gravity.

Man survived. Many species did not. Caves and crevices all over the planet are stilled jammed with a frozen mucky mixture, like a snapshot of bones and plants that early scientists incorrectly assumed never coexisted. Wooly rhinoceros, giant sharks, ancient rain forests, and modern humans all died on the same day, only 11,500 years ago. The Alps, once a line of lowlands, were pulled ten-thousand feet up and one hundred miles north of their original position.[4] Errant boulders weighing hundreds of tons were blasted across millions of acres by thunder so loud it could be heard on the other side of the world. Interplanetary methane, discharged with millions of volts of electricity formed long-chain hydrocarbons and rained upon the earth like pale milk. It was called Hoar Frost by the Israelites.[5] It was sweet to the taste and smelled like honey in the early hours of the morning. In the heat of the day it melted. Forward scouts sent by Moses returned with reports of land flowing with milk and honey. But, without stars and the sun to guide them, the Israelites only moved a few miles at a time. The sky was so low at times they had to lie on the ground to see any distance at all. Only a small fraction of these heavenly carbohydrates were consumed by animals. The vast majority seeped into the earth forming huge underground lakes and geysers, which today are only barely tapped by drilling a few miles into the earth. For nearly a century the earth has blossomed to unprecedented prosperity through the combustion of this raw form of energy, mistakenly called fossil fuel.

In a very short time, cosmically speaking, the earth was changed forever. Life spans were shortened. New symbiotic relationships would be formed. The plant and animal kingdoms had new neighbors. Entire populations died. New ones emerged in their new world. The very course of the earth was changed. For a hundred centuries man would try to explain and remember these events. That we should never forget, they made songs and scripture and legends. Whether random cosmic and catastrophic chance, or the actions of a mighty and intelligent command of God Himself can only be determined by the Spirit. One choice leaves only the despair of the pagan heart. The other spawns the realization of a family inheritance of galactic proportions, provided one is obedient to that creator. The choice belongs only to man. Choose now and carefully.

An apocalypse similar to this may be found in the *Old Testament*. The writings of Moses record that the Red Sea split in two and was pulled up into two walls of water, leaving dry ground where the sea once rested. While the collapsed sky rested on the Egyptians, and the ominous fire of the commanding light in the sky blazed into them, the Israelites were able to migrate across to the other side. This may have taken two or three days. And, what we now call a day—24 hours for a complete circuit of the sun—may have been 72 hours or more. Hours—more likely it took days for the house of Judah to walk across a distance of nearly 29 kilometers at its narrowest point and possibly as much as 90 miles—after the

Israelites were on the other side of the sea——the mists of darkness lifted from the Egyptians. Half emboldened from their wonder that they had not been hurt by this planetary assault, and enraged by their hatred for the Israelites, the *Old Testament* records they charged over the shallow shoals and across the bed of the sea. By this time the largest of these heavenly tormentors was moving away from earth. The waters of this shallow sea, now released from their gravitational maestro, crashed together like a mile-high tsunami and devoured the entire Egyptian army.[6]

The migrating Israelites finally stopped at the base of Mount Sinai, some 80 additional kilometers from the Red Sea crossing to the Southeast. Is it any wonder that without the commanding leadership of Moses, the anxious Israelites fashioned a golden likeness of the very comet/planet that delivered them from their dreaded enemy? Who's to say that these same forces, directed by a heavenly hand, did not work upon our planet in rather regular succession?

In fact recorded history seems to support this. Every 49 years—a week of seven years—a similar albeit smaller-scale event is recorded. Was it this errant planet that wreaked so much havoc on historical observers? Why was there a custom of freeing all slaves and forgiving all debts every 49 years called the year of Jubilee? Some scientists say it was Saturn. Some say it was Venus. In fact, the name *Venus* means hairy planet.[7] The peoples of the middle East were so anxious about this periodic conflagration because of the history of earthquakes, storms, floods, interplanetary lightening, and regional fires that seemed to fall out of the sky and devour stone, as well as volcanoes and collapsed sky conditions. Life in many forms was threatened every 49 years with heavenly disaster. From one Jubilee to another, the leaders did not want to have their society unable to flee to freedom, or perhaps more importantly, they did not want to appear before God in debt to man or the holder of a slave.

Then, perhaps as suddenly as it started, the visiting cosmic judgment ended. We can easily see Venus brightly and calmly shining in the Northern hemisphere like a silver jewel harmlessly guiding sailors across the sea. There is no recent physics that predicts another such calamity. Of course, every major telescope on or off the planet is contracted nearly every minute of the day and night to look for something other than near earth objects—NEO's. It usually takes a very long succession of nightly overexposed space photography of exactly the same coordinates to notice anything moving in space. Rarely do such explorer's have the freedom to just look around to see what may be nearby. The result could be that just such a cataclysm may occur again. Who could tell? And if the you think about it, who would tell if they knew? What could possibly be gained by telling? That's right. You figured it out. Sure makes one think twice about passing up that great deal on a telescope, doesn't it?

End Notes

[1] *"A Prayer of the Raising of the hand of Ishtar,"* in *Seven Tablets of Creation,* ed I.W. King

[2] H. Winckler, *Himmels- und Weltenbild der Babylonier* (1901) p. 43.

[3] http://scienceworld.wolfram.com/physics/RocheLimit.Fhtml

[4] *Cataclysm!: Compelling Evidence of a Cosmic Catastrophe in 9500 B.C.* by D.S. Allan & J.B. Delair, Bear & Co., a division of Inner Traditions International, Rochester, VT 05767 Copyright © 1995 & 1997 by D.S. Allan & J.B. Delair.

[5] *Exodus*

[6] et al *Exodus*

[7] *Worlds in* Collision Emmanuel Velikovsky

The Arrival of Noah

This is a continuation of the chapter, **The Garden of Eden.**
The waters of the flood are upon the spiritual creation and the Ark
of Noah is borne up and lifted from the earth. However, at this
point, the Ark was not yet under way because it was tethered to the
landscape until the ship had enough draft clearance to safely pass
over the submerged landscape.

In the Mesopotamian accounts, Noah (Utnapishtim), has
appointed the sailor, Buzur-Kurgala, to be in charge of the great ship.
We read in the account that "the cables of the ship were cast loose,"
and another sailor named Uragal parted the anchor cable. *Genesis*
7:18 reads, "And the waters prevailed, and were increased greatly
upon the earth; and the ark went upon the face of the waters." It was
not until the cables were loosed that the ark **went** or moved upon the
face of the waters. We read further in *Genesis* 7:20, "Fifteen cubits
upward did the waters prevail; and the mountains were covered."
The fifteen cubits upward is speaking about the draft of the vessel.

The ship laden with tonnage of supplies and all aboard, drew fifteen cubits of water which enabled the vessel to pass safely above all the obstructions of the landscape without scraping the bottom or running aground.[1]

David Fasold in his book, *The Ark of Noah,* states that drogue stones were used to restrain the ark from being cast sideways and broaching in a following sea if it didn't maintain a position keeping the sharp upswept stern into the waves. He further states that it was not the weight of the stones that accomplished this but the surface area of the stone. That's why they are long and flat and need not be thicker than what is needed for strength, not weight. The stones varied slightly in size but averaged 10 feet in length and 5 feet in width, with the thickest portion near the top by the hole where the lines were attached. Several of these stones may have trailed on longer lines far astern of the others in a forceful sea. There were at least fifteen tons of drogue stones trailing astern. They were never hoisted aboard, but in all likely hood had spares aboard to replenish a percentage of those lost.[2] The *Sin-Lequi-unninni* version of the *Epic of Gilgamesh* mentions a nautical three-stranded tow rope in Tablet V, column 11, and in Tablet X, column 11, there is found mention of drogue stones.

Urshanabi is the boatman who will take Gilgamesh across the waters of death to the dwelling place of Utnapishtim or Noah. In the account there has been some sort of incident and Urshanabi reprimands Gilgamesh for the damage he's done.

Line 37 "Your own hands, Gilgamesh, have hindered the crossing."

Line 40 "The stone things are broken."

Perhaps in the incident the drogue stones were stuck, breaking the holes out, and freeing their attaching lines. An argument ensued but apparently feelings were patched up as well as the equipment.

Line 41 "...and the stone things (he loaded) in the boat,"

Line 42 "(the stone things) without which (there is no crossing death's waters.)"

These drogue stones were most important to the ship's safety in the crossing. It can be compared as being afraid to drive a car without brakes as the drogue stones had brake like action. An anchor or a "dragging stone" was necessary to keep the boat from broaching and throwing those on deck into the waters of death. [3]

Herodotus relates the use of drogue stones by vessels sailing the Nile in Egypt. To prevent currents pushing vessels ashore on the bank when rounding bends, a low-lying raft was equipped with a sail under the raft, which caught the river currents full force. This was made fast to the ship's bow by a tow rope. Men on the raft then shot off downriver with the vessel trailing behind.

The raft was equipped with sweep oars and darted around the curve in the river with the boat's bow "in irons." The vessel's stern tended to turn around but was made to "track" through the use of a "braking stone" that weighed about 120 pounds. The weight of the stone applied just enough tension on the stern to make the tow follow correctly.[4]

The Book of Enoch the Prophet records, "In those days Noah saw that the earth became inclined (sinking), and that destruction approached. And he said, Tell me what is transacting upon earth; for the earth labours, and is violently shaken. Surely I shall perish with it." [5]

The Sanskrit text of the ***Mahabharata*** (Vana Parva) states, "The earth, as if oppressed with an excessive burden, sank down, suffering pain in her limbs, and the earth in distress was deprived of her senses by excessive pressure." [6]

The Babylonians described the earth's sinking as: "the god of the underworld tore out the posts of the dam." [7]

Rabbinical sources add that the earth was quaking amidst lightnings and thunderings, then a very loud terrible sound was heard in all the earth, never heard before. The sun became darkened and the foundations of the cosmos were dislodged.[8]

The Babylonian accounts continue as follows, "All the earth spirits leaped up with flaming torches, and with the brightness thereof they lit up the earth." [9]

Another Rabbinical source reveals that during the "seven days" the earth was flooded by sheets of light, and terrifying signs and commotion filled the heavens.[10]

The father of Methuselah, while lying down in the house of his grandfather, Malalel, saw in a vision the earth sinking into a great abyss to its destruction.[11]

The above accounts are descriptions of the spiritual earth being torn or removed from its foundation and orbit and the beginning of its descent into the blackness of space described in the above account as the great abyss. The sheets of light described above are the electrical discharges from the spiritual earth into the heavens as it begins it descent. The sky is blackened and described above as the sun becoming darkened but what really has happened is the spiritual earth, with its protective atmosphere, has left its orbit from the spiritual sun and is falling down into the blackness of space. The falling or sinking action into space exerts a feeling of tremendous pressure felt by all aboard the ark. The electrical storms are raging with tremendous waves from the "fountains of the great deep being broken up", and from "the windows of heaven that were opened." Chapter 6:28-31 of *The Book of Jasher* describes the intensity of the raging storms on the surface of the spiritual globe as follows: "and it (the ark) was tossed upon the waters so that all the living creatures within were turned about like pottage in a cauldron. And great anxiety seized all the living creatures that were in the ark, and the ark was like to be broken. And all the living creatures that were in the ark were terrified, and the lions roared, and the oxen lowed, and the wolves howled, and every living creature in the ark spoke and lamented in its own language, so that their voices reached to a great distance, and Noah and his sons cried and wept in their troubles; they were greatly afraid that they had reached the gates of death."

Lamentations 1:2 records this event, "...and cast down from heaven unto the earth the beauty of Israel (the beauty of Israel is the spiritual earth), and remembered not his footstool in the day of his anger!" *The Zohar* or *The Book of Splendor*, states, "He hath cast down the earth from the heavens." [12]

As the spiritual globe was descending through the spiritual universe where it was created, the rain and storms continued for forty days and forty nights. The forty days and forty nights describes only the length of time the rains continued. It is implying that for the equivalent of 40 days and nights the rain continued for 24 hours without stopping because the earth was now only experiencing darkness. Rabbinical Sources state that during the Deluge "the sun and the moon shed no light." [13] Other same source rabbinical writings say Noah and his crew were given "light" that illuminated all of the ark and served as a time-piece, to distinguish night from day, during the Fall. The source of the "light" came from *The Divine Book of Wisdom,* which was made of sapphires and housed in a golden casket. (*The Legends of the Jews*, Vol. 1, p. 142)

The spiritual or "airy" earth was rapidly descending through the various levels of the spiritual universes, traveling in the fourth dimension. At each level of entry into the different spiritual universes, heavens, or dimensions, the enormous mighty celestial Gates of Heaven are opened by the Watchers, who are surrounded by a flaming fire, and by the command of God, they let the spiritual or "airy" earth pass through its portals and continue its descent.

Charles L. Wlaker, a contemporary of Brigham Young, a Mormon prophet, heard the sermon containing the following statement. He reported in his diary July 13, 1862:

".... in the P.M. Brother Brigham spoke.... said that when this world was first made it was in close proximity to God. When man sinned it was hurled millions of miles away from its first position, and that was why it is called the Fall."[14]

As the spiritual or "airy" earth was falling rapidly through the spiritual universes or dimensions it was also passing through time zones. It was literally passing through millions of years in a matter of days. It may have had the appearance of a " shooting star" in its fall through the fourth dimension or maybe some sort of hyperdimensional space travel—discussed in *The Dimensions*— was the mode of transport. By what ever means of space travel was employed, the spiritual globe was traveling rapidly through time, space, and dimensions. In the meantime, the temporal earth (our planet) in the temporal or physical universe, was experiencing a terrible cataclysm that nearly destroyed the planet. Our solar system had a visitor that passed within the orbit of Jupiter and close to the orbit of earth. Its gravity combined with its speed and trajectory caused havoc to earth and other planets. The preceding chapter tells of this event and it is even chronicled in *Job* 38 verse 32: "Canst thou bring forth Mazzaroth in his season? Or canst thou guide Arcturus with his sons?"

The cosmic visitor approaches closer and the events on earth become increasingly larger. The tidal waves become enormous tsunamis of gigantic proportions. The gravity of the approaching interloper pulls the waters of the oceans out of their beds, attaining miles in height. When they are released from the pull of gravity they fall back as great crashing waves of immense power and destruction producing similar effects of a great flood, such as this planet has never before known in historical times. The temporal planet was experiencing a flood and destruction nearly the same time as the spiritual globe. *Psalms* 33:7 describes this event: "He gathereth the waters of the sea together as an heap: he layeth up the depth in storehouses." *Psalms* 93:3, "The floods have lifted up, O LORD, the floods have lifted up their voice; the floods lift up their waves." And *Psalms* 60:2, "Thou hast made the earth to tremble; thou hast broken it: heal the breaches thereof; for it shaketh."

The gravitational pull from the near collision of the cosmic interloper caused terrible earthquakes, volcanic eruptions, raging tempests, and fracturing of the planets crust that resulted in the earth being slowed by the dragging of the crust as it was pulled off. The Earth slowed its rotational speed, and the distance of the Moon from the Earth was changed—some say closer and others say further away. Earth lost approximately 10% of its circumference and more than ½ of the crust. This caused the tilting of the planet's axis to a new angle and created the wobble in its new axis. The oscillations in a back and forth surging motion caused continents to move to new locations. Mountains rose in minutes. *Psalms* 114:4 states, "The mountains skipped like rams, and the little hills like lambs." The continents in their move left marks or tracks in their wakes that solidified. You can see the tracks that India left as it raced towards the Asian continent. South America can be moved back into Africa and Australia into Antarctica by tracing the tracks that are carved in the basins.

Most of the continents had changes in terrain as they smashed into each other resulting in folding and bunching up. Millions of years of uniformitarianism continental drift were accomplished in days.

The planet was left wounded with a gaping hole on one side and the remaining crust on the opposite side. For up to twenty-five years or more , the planet suffered a collapsed sky, darkness and unbearable cold. Ninety percent of life was wiped out. It was when the planet nearly died 11,500 years ago or 9,500 B.C.[15]

Immanuel Velikovsky's book, *WORLDS IN COLLISION*, recognized the compelling evidence of a catastrophic world event caused by a runaway planet that nearly collided with earth; a planet that he identified and thought was Venus, now in a stable orbit. He recognized the cataclysm, but wrongly identified the cosmic interloper.

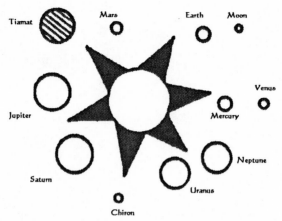

Babylonian cylinder-seal, which includes Tiamat and Chiron. We consider Chiron an asteroid, but even so, the 11-planet system includes one more planet than our current solar system with Tiamat missing.

There is an ancient Babylonian cylinder-seal in the East Berlin State Museum that apparently shows the planets in more or less their correct order, in an anti-clockwise direction or in the direction that they naturally orbit the sun. One planet identified as Tiamet is missing. What happened to this planet? There is an ancient Akkadian creation epic that describes a "war in heaven" between the gods (planets being named for gods) that tells of a planet (Marduk alias Phaeton) entering into the solar system from the heart of the Deep (cosmic space) where he was created.

Tracing by R. Verrill of part of Akkadian cylinder-seal VA/243 in State Museum, Berlin. Note the same 11-planet system and the difference in height between the seated fugure and the standing figure.

The epic states:

"In the Chamber of Fates, the place of Destinies, A god was engendered, most able and wisest of gods; In the heart of the Deep was Marduk created."

The saga describes Marduk as a radiant and huge visitor from a distant interstellar region spewing great jets of fire.

"Alluring was his figure, sparkling the lift of his eyes; Lordly was his gait, commanding as of olden times...Greatly exalted was he above the gods, exceeding throughout…When he moved his lips fire blazed forth."

After entering the solar system, Marduk first encountered Neptune or Ea, then Uranus or Anu, passing Saturn or Anshar but became trapped for a time by the three large outer planet's gravitational fields. This began to affect the orbit of Tiamet and its satellite Kingu which appeared to earth's peoples as an onslaught of gods preparing for war.

Marduk plunged on and began to approach Tiamet and the inner planets. The epic then tells how Tiamet under attack from deranged Marduk begins to break-up in a celestial battle as follows:

"Tiamet and Marduk, the wisest of the gods,
Advanced against one another;
They pressed on to single combat,
They approached for battle."

Marduk then:

"…spread out his net to enfold her;
The Evil Wind, the rearmost, he unleashed at her
face. As she opened her mouth, Tiamet, to devour
him- He drove the Evil Wind so that she closed not
her lips. The fierce storm Winds then charged her

belly; Her body became distended; her mouth had opened wide. He shot there through an arrow, it tore her belly; It cut through her insides, tore into her womb. Having thus subdued her, her life-breath he extinguished."

Marduk, having demolished Tiamet, drags the remains of Tiamet with it forming a "tail" of debris which streams out behind as a "Great Band" or "bracelet." Today these probably represent the asteroid belt.

Jewish scripture, refers to Marduk simply as "the Lord":

"The Heavens bespeak the glory of the Lord,
The Hammered Bracelet proclaims his handi-work...."

And very significantly adds:

"From the end of heaven he emanates."

As already noted from the Babylonian epic, Marduk came from the "Great Deep" or end of heaven. Marduk then continued wandering erratically among the planets in a sunwards direction, narrowly missing earth but causing conflagration, tectonic upheavals, crust removal, and a tremendous flood on this planet. It is believed that Marduk or Phaeton plunged into the sun. What if it didn't? Will this cosmic visitor one day return to again reek havoc in our solar system?

Outline drawing in W. Hayes Ward of an
apparent 'planetary' system on Bablyonian
cylinder-seal VA/508

More evidence is found in the Indo-Iranian *Zend-Avesta*, as these planets:

"...ran against the sky and created confusion.", and "The celestial sphere was in revolution....The planets, with many demons, dashed against the celestial sphere and mixed the constellations." These "demons" were presumably the debris of Tiamet.

Homer's *ILLIAD*, contains a lengthy account of a celestial battle. The story is pretty much the same but with Greek names of gods and goddesses attached to the planets.

Paraphrasing Homer's account we find that Pallas-Athene (Venus) removes Ares (Mars) from the battle field. (Mars orbit is altered), only for the two to meet again-this time "furious Ares abiding on the left of the battle." (Marduk passes around behind Mars and crosses inside Mars' orbit) Aphrodite (the moon) tries to enter the conflict. (the Moon affected gravitationally) Ares then darkens the battlefield (the Martian atmosphere becomes disturbed) and Pallas-Athene is seen to retreat. (Marduk moves away from Mars). Hera, the goddess of Earth then complains at this celestial conflict. (Earth is affected electromagnetically) Numerous lightening bolts described as "spears" are exchanged between Pallas-Athene and Ares. Pallas-Athene retaliates by heaving a gigantic "black, jagged, and great stone" (the remaining portion of the dead Tiamet) at Ares and thereby "loosed his limbs." (caused Mars to oscillate). Aphrodite (the moon) came to Ares assistance, whose "hand" she took and "sought to lead away." (the moons gravitational pull caused Mars to move further away from Marduk); but Pallas-Athene "sped in pursuit and smite Aphrodite on the breast...and her heart melted." (Marduk discharged an electromagnetic bolt at the Moon).

Earthbound observers perceived Venus attacking Mars but it was something bright and star-like (Marduk) that became embroiled in "heavenly strife" with Mars.

Old Persian texts expressly state that Tistrya, the " leader of the stars against the planets", proceeded "along his winding course." Phaeton pursued a meandering sunward course through the solar system from planet to planet much like playing pin-ball.[16]

These ancient writings, sagas and epics recorded the cataclysm that nearly destroyed the earth and the conflict in the heavens, long thought to be myths, but written to the ancient peoples understanding and interpretation of the events that took place.

In the meanwhile, the spirit or "airy" Earth has passed through the last portal of heaven into the temporal universe and descends to its "soul mate" and carbon copy, the temporal planet. It has taken a year of celestial time to Fall through the Heavens. It immediately attaches itself and enters the temporal planet. The two planets have united and become as one. *The Zohar* affirms this unity in the following writing, "this is the supernal (spiritual) world to which this earth is attached and from which it is nourished."[17] The word nourished, used in context here, would mean replenished. The waters of the flood are now upon the temporal planet but not one drop of water has wet this planet because it was spiritual water. This event occurred some time later after the cataclysm because of time travel through the fourth dimension. When the spirit earth reached its destination, several thousand years had passed since the temporal planet had experienced a near death blow.

The Zohar comments on the cataclysm and spiritual earth being removed from heaven as follows:

"And when the Holy One resolved to destroy His house and the Holy Land below,

He first removed the "Holy Land" which is above—the celestial prototype—and

Cast it down from that grade where it had formerly imbibed nourishment from the holy Heaven, and then **He caused the land below to be devasted:** first **He cast down from heaven the earth**, and then He remembered not his footstool"[18] (Bold type added)

It wasn't until the two planets united that this planet was called Earth. The words of the song, *Colors of the Wind*, from the movie, *Pocahontas*, so movingly expresses the Redman's knowledge of this event as:

"You think you own what-ever land you land on;
the earth is just a dead thing you can claim;
but I know every rock and tree and creature has a
life, has a spirit, has a name."

The temporal planet now has a spirit and a name, EARTH. The earth, with all its animals, the fishes of the sea, and the fowls of the air, have become living souls.

Latter day revelation reveals that the Earth, the animals, the fishes of the sea, the fowls of the air, as well as man, are to be recreated, or renewed, through the resurrection, for they too are living souls.[19]

The Zohar supports this belief by the following writing: "Let the earth bring forth a living soul according to its kind, namely, cattle, and creeping thing, and beast of the earth after its kind."[20]

Back to Noah. The first drogue or "signal" stone that touched bottom gave warning that the Ark was coming into shallow water. Noah knew that he had to act quickly to determine if the shallow water was a suitable landing site because receding waters cause strong raging currents that could drag the Ark over dangerous landscapes to its destruction.

Acting quickly while the tidal bulges were still in effect, he unlashed and lowered the steering oar to below the bottom of the hull. Drogue stones were positioned on either side of the Ark to help guide and hold position, and one drogue was positioned aft to act as a braking stone should the current prove too strong. The Ark continued its forward movement, gliding, gently eastward upon the receding tides toward the slopes of Mount Mahser, when Noah called to have the shaft of the sweep oar rotated, slowly causing the Ark to veer in a semicircle toward the south, where the "High Place of Cudi" (Al Judi) acted as a point to check against sideways drift.

The "braking drogue stone" touched bottom surging the line taut, slowing the great ship, and now and then catching on obstructions below further slowing its forward movement. The starboard drogue stones were slackened to decrease the angular track, and the great Ark came back down to earth settling on the soft slopes of Al Judi buckling the large steering oar under the weight of the hull.[21]

The Bible says, "And the Ark rested…. upon the mountains of Ararat", but in the original King James Version Ararat was rendered as Armenia. The *Qur'an* in Hud 11:44 states, "…and the ark rested on the Judi." According to Moslem tradition the Ark is on a hill, not a mountain, but upon the mountainous range of Urartu, (Ararat or ancient Armenia) and from the *Qur'an*, Al-Judi. The following writings of Berosus seem to make it most clear that the landing site was in Armenia:

"A portion of the ship, which came to rest in Armenia, still remains in the mountains of the Kordualans of Armenia, and some of the people, scraping off pieces of bitumen from the ship, bring them back and use them as talismans."

Mount Ararat has a deep chasm or rift known as the Ahora Gorge, caused by an earthquake in 1840. In the mountain of Ararat and the Ahora Gorge there is a piece of ice in the shape of an upside-down heart. This area is the peak of the mountain known as Al Judi.

The Moslem *Qur'an* states that the ark came to rest on Al Judi, and the people for many years thought Al Judi was a completely different mountain. Recently it was learned that Al Judi refers to this particular peak in the mountain of Ararat. Therefore, it appears that the *Bible* and the *Qur'an* refer to the same mountain.[22] This was probably Armenia in ancient times.

Genesis 8:13, 15, 19 states,

"And it came to pass in the six hundredth and first year, in the first month, the first day of the month, the waters were dried up from off the earth; and-Noah removed the covering of the ark, and looked, and, behold, the face of the ground was dry.

And God spake unto Noah saying, "Go forth of the ark, thou, and thy wife, And thy sons, and thy sons' wives with thee.

Every beast, every creeping thing, and every fowl, and whatsoever creepeth upon the earth, after their kinds, went forth out of the ark."

The Lord having dried off the spiritual waters from the face of the temporal earth now commanded Noah and all those souls and animals aboard the ark to disembark. Once all those souls on board, including the animals and fowls, touched the dry land, they became fully mortal and the ark became temporal. The invisible became visible. In the twinkling of an eye their celestial nature was changed to that of the earthly without even being aware that such a change had happened. *The Zohar* states that spiritual beings in descending on earth must put on themselves earthly garments (flesh and blood), as otherwise they could not stay in this world, nor could the world endure them.[23] It happened on the very birthday of the patriarch, or the 601[24] year, first month, first day of the month, of the life of Noah; bearing in mind that this was still a celestial year and not measured the same as on their new place of habitation. *The Zohar* implies that the length of a celestial year could be a much longer time than currently thought by the following statement: "And God remembered Noah," for Noah having been so long out of sight had to be specially brought to mind.[25]

Noah knew that he was in a different place because the light of the sun and moon was noticeably not as bright.[26] The order of nature was reversed, the sun now rising in the east and setting in the west, instead of rising in the west and setting in the east. The sky was different.[27]

Another change was that this planet had seasons or a time in which to plant and harvest crops and climates of hot and cold areas, due to its new axis tilt. The spiritual earth apparently was always temperate with a vertical axis. The Lord revealed these changes to Noah in *Genesis* 8:22:

> "While the earth remaineth, seedtime and harvest, and cold and heat, and summer and winter, and day and night shall not cease."

Most noticeably was the dietary change. The Lord told Noah "that every moving thing that liveth shall be meat for you". Prior to the deluge all were vegetarians, including the animals. Apparently temporal bodies of both men and animals require meat for health. Some of the animals that were on the ark became carnivorous. All the animals and fowls, once tame, now feared and dreaded Noah and his family.

Genesis 9:1, 2 states:

> "And God blessed Noah and his sons, and said unto them, Be fruitful, And multiply, and **replenish** the earth."

> "And the fear of you and the dread of you shall be upon every beast of the earth, and upon every fowl of the air, upon all that moveth upon the earth, and upon all the fishes of the sea; into your hand are they delivered."

Another new phenomenon was rain. There was no rain on the spiritual earth until the flood when "the windows of heaven were opened." Now the Lord gives a rainbow as a token of a covenant between him, Noah, and life on the temporal earth as stated in *Genesis* 9:14 &15 as follows:

> "And it shall come to pass, when I bring a cloud over the earth, that the bow shall be seen in the cloud:
>
> And I will remember my covenant, which is between me and you and every living creature of all flesh; and the waters shall no more become a flood to destroy all flesh."

Rain has been on the temporal earth almost since its beginning but unknown on the spiritual earth. Everyone here knows that rainbows are formed opposite the sun by refraction and reflection of the sun's rays in rain, spray, and mist, producing a bow or arc that exhibit the colors of the spectrum. It must have been a beautiful and impressive event as seen by Noah and his people for the first time.

Another change, although not immediately noticed, was mortality. The life span of man was gradually shortened until the days of Moses, and then fixed at 120 years maximum. The ancients did attain a long duration of life because they were beloved by God because of their virtue and the good use they made of it in astronomical and geometrical discoveries, which would not have afforded the time of foretelling the periods of the stars unless they had lived six hundred years; for the great year is completed in that interval. Josephus further states in his writings that the ancients lived a thousand years, which is verified by many ancient historians such as Manetho, Berosus, Mochus, Hestiaeus, Hieronymus, Hesiod, Hecataeus, Hellanicus, Acusilaus, Ephorus and Nicolaus.[28] In other words the ancients were permitted a longer duration of life in order to preserve knowledge. Noah lived nine hundred and fifty years and died a mortals death without being translated as previously most had been taken or translated at death on the spiritual earth. Again, it must be emphasized that a year of temporal earth time was not the same as on the spiritual earth that used celestial time. It took the ancients a long time to rethink time on this sphere as to hours, days, months, and as to what constituted a year and at the same time studying the movements of astral bodies in this solar system. Until time was recalculated celestial time was continued. Noah may have lived much longer than 950 years if celestial time is greater than temporal earth time.

The temporal earth having passed through a great cataclysm several thousand years previously had need of replenishing that which was lost. The animals, from the ark, paired off and eventually joined other animal populations that survived the cataclysm. There may not have been giraffes, monkeys, skunks, etc. on board the ark as these animals could have survived the cataclysm in sufficient numbers to replenish themselves. Only the animals, fowls, and creeping things **from** the spiritual creation were on board the ark. There were no noxious insects on the ark. Insects were only on the temporal earth. The spiritual earth was free of flies, wasps, hornets, ants, beetles, mosquitoes, etc. We guess that is why it was considered a paradise. Creeping animals refer to small slow animals such as sloths and opossums. Turtles could have survived the cataclysm by simply digging in. We don't know for sure what kinds of animals came from that creation but they were brought here to replenish and, to add to the diversity of earth's existing animal and fowl life forms that had survived the great cataclysm.

"This is why the ark didn't have to be the size of the state of California, my dear daughter, Eve." This is the answer to the question that my daughter asked me about 15 years ago. Refer back to the Forward section of this book for the question.

Many new modern, smaller size life forms of animals made their appearance on the scene suddenly, all arriving on the ark. Ancient species such as the woolly mammoth, giant saber tooth tiger, giant sloths, huge armadillos, giant anteaters, and other species of giant size, were extinguished in the cataclysm. Maybe the newer, more modern, smaller size animal species such as the Indian elephant, tigers, armadillos, and anteaters, were on the ark as their replacement. One creature that probably arrived on the ark was the plain domestic house cat. From earliest recorded times, the domestic house cat has preferred to make its home with man, and always domestic. They are only wild if forced to fend for themselves without human contact and even then are very easily tamed. Another animal that has a different genetic background is the dog. Wolves were common in large and small species on the temporal earth. The modern species of dog did not descend from the wolf. Although there have been instances where wolves were successfully bred with dogs, the genetic code from dogs could not have been derived from wolves. The modern dog was more likely introduced to the current race of man from the ark. We do know from the *Bible* scriptures that cattle, doves and ravens were on board and that both clean and unclean animals were taken aboard. Noah built an altar unto the Lord; and took of every clean beast, and of every clean fowl, and offered them as burnt offerings on the altar. The clean beasts were probably sheep, cattle, and goats, and the clean fowl were probably doves.

Noah afterwards became a husbandman or breeder of domestic animals, most probably cattle, sheep and goats. Maybe some horses and a few camels. Perhaps chickens, ducks and geese. He planted vineyards and perhaps introduced new varieties of grains.[29]

This chapter will be continued in the chapter titled, *The Brave New World.*

The flood was indeed a global flood. It happened on the spiritual earth, which united with the temporal earth and became accounted for as one planet, now called Earth. Therefore, in a sense, the flood occurred on both the spiritual and temporal earths. It is the same for us as we have a spiritual soul and temporal body that have united but are accounted for as one being or personage. Because both planets are now one, the history that applied to the spiritual creation applies to the temporal earth as well. In this enlightened view the flood indeed occurred on Earth and it was global. This union is called the union of the polarity and will be discussed in the following chapter.

The Zohar IV explains the verse "In the beginning God created the heaven and the earth", means that the lower world was created after the pattern of the upper. (p. 289) What The Zohar is trying to say in the above scripture is, "In the beginning God created the heaven (upper or spirit earth) and the earth (lower or temporal earth). When one views the above verse "enlightened" by ancient Jewish interpretation, the meaning becomes much more profound and clear about the creation. *Genesis* now becomes understandable when it implies that there were two creations. The word *heaven,* meant the upper or *spirit earth* to the ancients. Terminology and meaning of words can change with the ages, even to the meaning of the simple word *heaven.* Had we retained the ancient meaning of *heaven,* as used in *Genesis,* it may have not been necessary to write this book as the religious theologians would probably have figured it all out by now.

Down through the ages the 9,500 B.C., cataclysm of the temporal earth and the flood of Noah story have become confused and intertwined. The two events were separate and occurred several thousand years apart, yet close enough in time to be thought of by the descendants of the ancients as the same event. It is this confusion of the two events that have given rise to so many myths and flood stories that are similar to one another yet, each contain some truths.

Noah when he arrived saw the destruction on the recovering earth from the cataclysm and wrongly assumed it was from the deluge, thus the source of the two events becoming entangled.

The Mesopotamians regarded Noah and his family as gods because they came through the judgment and because of the longevity they carried with them into the new world.[30] The story of how Noah and his family survived the deluge by traveling from the heavens to earth by sailing across the cosmic sea in " the divine boat of Noah" or " The Boat of a Million Years" became legendary, adding to their status as gods and heroes. The cosmic sea is the dark rift in the Milky Way that contains a black hole. Egyptian legends suggest that Noah's divine ship sailed through the black hole into this universe. Of course, this would be the spirit earth traveling through a black hole transdimensional portal. (see chapter *The Dimensions*)

The Egyptians were notorious "borrowers"and modifiers of legends and religion to serve the political scenario of the times. The Egyptian *Book of the Dead*, chapter clxxxv, by the scribe Ani, refers to the flood story. Referring to this account E.A. Wallis Budge, late keeper of Egyptian Antiquities at the British Museum, informs us in his work, *The Egyptian Book of the Dead* as follows:

"A general destruction of mankind was caused by the Flood which was brought upon the world by the god Temu, who announced his intention of destroying everything in it and of covering the earth with the waters of the primeval ocean Nu. All life was destroyed, and the only beings who survived were those who were in the 'Boat of the Millions of Years', i.e., the Ark of the Sun-god, with the god Temu."[31]

"The Egyptians replaced Noah and his family with their god, Ra, and his companions whose Sun Barque sailed over the heavens daily but originally it was known as the Uaa Nu, "the divine boat of Nu," or Noah. Shem was known to the Egyptians as one of the eight survivors of the flood and one of the godlike antediluvians from "the divine boat of Nu."[32] When the gods use the great ship to travel between the worlds, it is called "The Boat of a Million Years."

Again the *Papyrus of Ani* in Plate XX11, line 20, states that the boat of Nu is said to be the color of green for the divine chiefs.

The authors wish to restore Noah and his family back to their rightful place and claim to fame as survivors on board the "divine boat of Noah." And since it traveled between the great universes through the fourth dimension of time and space deem it appropriate to call his divine ship, "The Ark of Millions of Years."

".... by the word of God the heavens were of old, and the earth standing out of the water and in the water:

Whereby, the world that then was, being overflowed with water, perished:"

"But the heavens and the earth, which are now, by the same word are kept In store, reserved unto fire against the day of judgment and perdition of ungodly men." *II. Peter* 3:5, 6, 7.

Could it be that the above scripture of the earth standing out and in the water, apparently at the same time, refers to when the flooded spiritual earth merged with the temporal earth? And the waters of the flood were on both planets but not a drop of water wet the temporal earth because it was spiritual water? The earth, in a sense, was standing in and out of the water at the same time.

Does not the scripture refer to a "new heaven and a new earth" (the temporal earth) which are now, the old heavens or world then (the spirit earth having left its former abode) having perished?

It took from the time of Adam's transgression to the flood of Noah and the destruction of the spiritual earth for the Fall to be complete.

E. J. Clark & B. Alexander Agnew, PhD

End Notes

[1] (*The Ark of Noah, David Fasold, 1998, Knightsbridge Publishing Company,* NY, NY, p. 166)

[2] (*The Ark of Noah,* pp.170,171,177)

[3] (*The Ark of Noah,* pp.180-181)

[4] (*The Ark of Noah,* pp.170-171)

[5] Excerpts taken from *The Book of Enoch The Prophet,* ©2000. Translated by Richard Laurence. Published by Adventures Unlimited Press, Kempton, ILL. All rights reserved. Used by permission of the publisher., p. 78)

[6] (The Ark of Noah, p. 163)

[7] (et al p. 164)

[8] (*Sefer Hayashar, The Book of the Righteous,* Ha-Yewani Zerahiah and translated by S. J.Cohen (NY, 1973)

[9] (Ref; *The Ark of Noah,* p. 164)

[10] (Tractat Sanhedrin 108B of the *Babylonian Talmud*)

[11] (*The Book of Enoch the Prophet,* et al pp.118, 119)

[12] (*The Zohar,* translated by Simon, Sperling and Levertoff, ©1984, London, The Soncino Press, Ltd., Vol. 11, p. 311)

[13] (*The Legends of the Jews,* L. Ginzberg, Philadelphia, 1928, vol. 1, p. 162)

[14] (Diary of Charles Lowell Walker, 2 Vols. , A Karl and Katharine Miles Larsen, eds., Logan: Utah State University Press, 1980.)

[15] (*Cataclysm!: Compelling Evidence of a Cosmic Catastrophe in 9500 B.C.* by D.S. Allan & J.B. Delair, Bear & Co., a division of Inner Traditions International, Rochester, VT 05767 Copyright © 1995 & 1997 by D.S. Allan & J.B. Delair. back cover of book)(www.kamron.com, Earths Great Cataclysm, viewed 2/20/02)

[16] (*Cataclysm!: Compelling Evidence of a Cosmic Catastrophe in 9500 B.C.* by D.S. Allan & J.B. Delair, Bear & Co., a division of Inner Traditions International, Rochester, VT 05767 Copyright © 1995 & 1997 by D.S. Allan & J.B. Delair. pp.219-232)

[17] (*The Zohar* et al Vol. IV, p. 358)

[18] (*The Zohar* et al Vol. IV, p. 107)

[19] (Conference, Rep. , Oct. 1928, pp.99-100)

[20] (*The Zohar* et al 1, p. 56)

[21] (*The Ark of Noah,* pp.174-177)

[22] (www.parentcompany.com/search_for_noahs_ark/sfna1.htm viewed Nov. 30, 2002)

[23] . (*The Zohar* et al Vol. V, p. 211)

[24] (*The Zohar,* translated by Simon, Sperling and Levertoff, ©1984, London, The Soncino Press, Ltd., Vol. 1 p. 211)

[25] (*The Zohar* et al Vol. 1, p. 237)

[26] (Tractat Sanhedrin 108B of the *Babylonian Talmud*)

[27] (Taken from **The New Complete Works of Josephus** © **1999** by **Kregel Publications**. Published by Kregel Publications, Grand Rapids, MI. Used by permission of the publisher. All rights reserved , Book 1, Chapter 3, p. 55)

[28] (KJV *Genesis* 9:20)

[29] (*The Ark of Noah*, p. 101)

[30] *The Egyptian Book of The Dead,* translated by E.A. Wallis Budge, 1st published in 1967, Dover Publications, Inc. NY.

[31] (*The Ark of Noah*, pp.136, 194)

The Union of the Polarity

The union of spirit and matter is the union of the polarity. It is the union of opposing principles, later described as male and female counterparts, but in the beginning it meant only the union of spirit and matter, as when the spirit earth united with the temporal earth, and became one united sphere. In theory, when the union of the polarity occurs balance is achieved and where there is balance there is harmony.

"He founded the earth on its base"[1] This verse indicates that everything in the physical world has a specific spiritual counterpart and basis, through which it can be elevated. Spiritual things can be bound to the material, as, for example, the soul is bound to the body.[2]

The Union of the Polarity

Noah and his sons probably understood the dramatic event of the merging of the two earths, but later generations viewed the knowledge of the merging of the two earths as a "great mystery" and tried to explain the event by developing rites, philosophies, symbolism, and various interpretation of ancient writings to their understanding. An example is the following:

The spiritual earth (male principle) united with the temporal earth (female principle). The children of the East defined the male and female principles as the Yin (female principle) and Yang (male principle). Gradually these principles were applied to all opposing things such as hot and cold, life and death, sweet and sour, black and white, day and night, good and evil, etc. Somewhere, along the way, in trying to understand the union of the two earths, the concept of the Yin and Yang replaced the ancient knowledge and was applied to nearly everything except the union of the two planets. Another possible explanation is that the knowledge was generally accepted as true at first but gradually lost its credibility due to lack of understanding of the "great mystery" which they couldn't explain. In other words the event became more like a myth and from the myth sprang various forms of corrupted belief and doctrines.

Yin and Yang Symbol

Fortunately *The Zohar* books retained many bits and pieces of the union of the two planets and even by admission of various rabbis, poorly understood. They too, viewed it as a "great divine mystery." Below are a few of the many documentations found in *The Zohar* or *Book of Splendor*, to the union of the polarity.

"However, when the Holy One, blessed be He, created this world, He created it after the pattern of the supernal (spiritual world)

world. All the aspects of the upper world he established in the lower, so that the two worlds should be firmly knitted together." [2]

"The blue employed in the work of the Jewish Tabernacle symbolized the mystery of the upper world, the blue and the purple together symbolized the knitting together of the upper world and the lower." [3]

"And they bound the breastplate by the rings thereof unto the rings of the ephod with a thread of blue. Why blue? Because it is an all-uniting color, and thus is symbolic of the supernal (spiritual) mystery. (the mystery of the two earths uniting)[4] And, observe that all these measurements prescribed for this world had for their object, the establishment of this world after the pattern of the upper world, so that the two should be knit together into one mystery." [5]

"When the worlds were all completed they were joined into a single organism, and are symbolized by the letter Vau or the number six when it is united with the veiled world. (the temporal earth) [6] Earth is a copy of Heaven. Heaven is a copy of earth. They are no duality, but an absolute unity. [7] Because the upper world is called "great", the lower world, which is united with it, is also called "great"...."[8] Words in parenthesis added.

"And as Jacob and his sons proclaimed the union of the world above and the world below, so also must we. Blessed is he who concentrates his mind and will with true humility and longing upon this mystery." Said Rabbi Hamnuna, the Ancient: "This stirring up of the unity has indeed been rightly and justly expounded, and that which we have just now heard is indeed very true; and in the future time these words which we have now uttered will stand before the Ancient of Days and in no wise be abashed." [9]

In *The Book of Creation* or the *Sefer Yetzirah*, is found the ancient Kabbalah doctrine of the universe of Asiyah (Making), which consists of the physical world and its spiritual shadow.... "earth with its spiritual shadow". [10]

A detailed account of the creation is given in the Mayan *Book of the Month* of the *Chilam Balam*. The *Book of the Month* section is measured in named months. In the month of *Four IV* it states that "heaven and earth embraced each other." The heaven is the upper or spirit earth. When it embraced the lower or temporal earth, it was the Mayan way of expressing how the two earths merged or united into one sphere.

At various sacred sites in the Andes an ancient ritual is practiced by Shaman that continues today called *The Sacred Marriage of Heaven and Earth Rite*. The union of the two earths is celebrated in this mystical rite. In ancient Sumer (Babylon), as early as 3,000 B.C., this same ritual was practiced, called *The Sacred Marriage Rite,* initially as the union of the two earths, and later became corrupted. It was viewed as a marriage and sexual union between the male (spiritual earth) and female (temporal earth). Later everyone got into the act, gods and goddesses, priests and priestess', fertility rites, etc.

The Star of David is an ancient Jewish symbol representing the union of the two earths by its six points. See above about the number six and letter Vau.[11] The number 6 was considered sacred by the ancients. The Egyptians used a numbering system based on the sacred number 6, rather than the number 10, as we use today. Stonehenge was built using the same system and the Mormons also used the same numbering system to build some of their temples, especially the great temple in Salt Lake City. The sacred number 6 represented polarity union.[12]

The temple of Solomon incorporated strategically placed metal rods into its design. These rods united water channels (representing the female principle), which ran under the temple, to golden spires (representing the male principle) on the temple roof, creating a union of the polarity which they believed kept the two planets permanently fused together. The columns of the tabernacle symbolized the united Heaven and Earth (the roof and floor).[13]

It seems that by the time of Solomon the mystery of how the two earths had merged deepened and later Jewish rabbinical writings reveal the belief that the two planets would separate and reunite from time to time.

To ensure that every facet of the tabernacle promoted balance by the union of the polarity, Moses hired a representative of the tribe of Dan whose symbol was the serpent (female principle) and another representative from the tribe of Judah, whose symbol was the solar lion (male principle) to oversee the temple's construction.[14]

The pre-Christian cross was known among the ancients as the key to the mysteries of antiquity. It was always associated with water; the flood. To the Babylonians it was the emblem of the water gods. The cross symbol represented the mystery of the merging of the two earths. Anciently it was viewed as a "great mystery" as to how this event happened. Now the authors want to "unveil" this great mystery through interpretation of pre-Christian cross symbolism that crept into Christian cathedrals, government, religion, and architecture that testify silently of the event.

The Knights Templars, while searching for the "flying scrolls" in the ruins of Solomon's Temple, apparently uncovered the ancient knowledge of the union of the two earths by accident, resulting in an explosion of cathedral building throughout Europe that symbolically recorded the polarity union by employing the principles of sacred geometry.

While adopting a circular design for their churches, along with a perfect cube for an inner altar, the Templars symbolically represented the union of the spiritual earth (male principle) represented by the cube and the temporal earth (female principle) represented by the circle. Many of their churches and cathedrals were also built in the shape of the cross that also incorporated circular stained glass windows and circular building design in combination together. These buildings reflected the "pure" embodiments of symmetry, balance and proportion. The cross is an ancient symbol of the union of the polarity prior to the advent of Christianity.[15] Why did the Templars use so much symbolism? It was because they were oppressed by the Vatican's reign of terror known as the Inquisition. To speak out with unsanctioned church doctrines was an invitation to be subjected to torture, terror and execution.[16]

In Portugal, the Knights Templars, under persecution, reorganized and changed their name to the Knights of Christ. Vaso da Gamma was a Knight of Christ. Prince Henry the Navigator was a Grand Master of the order. The Knights of Christ sailed in ships under the Templars familiar red Maltese cross on their sails, the same cross that embellished the sails of Columbus's three caravels that crossed the Atlantic to the new world. These crosses were symbols of the union of the polarity, ancient knowledge found by the Templars.

Columbus himself was married to the daughter of a Grand Master of the Order and had access to his father-in-law's charts and maps that covered the Mediterranean area, and European Atlantic Coast, and other ancient sea charts that no doubt aided his voyage to America.

Not long after the Templars dispersal, very accurate and inexplicable sea charts began to appear all over Europe. These maps were far superior to the Ptolemaic maps that were in current use. It is thought that Phillip the Fair gained access to some of these maps when he seized part of the Templar's wealth in a surprise raid. These maps and charters were also part of the unexpected "find" the Templars found in the ruins of Solomon's Temple. Besides finding the maps and charters, the Knights Templars found information on map making, and how to improve navigational instruments such as the astrolabe and the compass.

Columbus's voyage was financed by a mysterious consortium of wealthy men (Templars) and Jews and not by the sale of Isabella's jewels as so commonly thought. Columbus was a Spanish Jew, so it was not mere coincidence that he weighed anchor on August 3, 1492, just a few hours before the deadline for all Jews to either convert to Christianity or leave Spain.[17]

Why did the Templars and Jews finance Columbus' voyage? They were not seeking more wealth or new trade routes as history would have us to believe but were looking for another "find" foretold in the "flying scrolls" they found in the ruins of Solomon's Temple. These manuscripts prophesied that a land existed across the ocean to the west whose destiny was to one day become the New Jerusalem.[18] Columbus's mission was to find this new land so that the Templars and Jews could found the New Jerusalem and create a "New World Order" based on the concepts of democracy and freedom. Nearly every charter to the new world discovered by Columbus required that all passengers sign a compact or submit to the terms of a charter to do everything in their power to spread the gospel of Jesus Christ. Many of these passengers sold everything they had, including their servitude and the freedom of their children for a fixed period of time, in order to be able to do one simple thing; practice their religion according to their own conscience.

The trident is a modified version of the cross. It represents the union of the male and female principles, symbolic of the union of the two earths, and the three powers of God or the Godhead, symbolized by the three prongs.[19]

In Mesoamerica the Tau Cross was the symbol of polarity union and associated with the "plumed Serpent" called Quetzalcoatl or Kukulcan and another serpent called Itzamna.[20] The Tau Cross was to be seen adorning statues and bas-reliefs at Palenque, Copan, and throughout Central America and parts of Mexico.

In the Mayan ruins of Uxmal, Yucatan, Mexico, the symbol of the Star of David was found with a feathered or pennated tail that suggests a common Assyrian, Phoenician or Persian origin. Uxmal dates about 1000 A.D. Below are two illustrations of comparison.

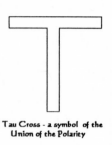

Tau Cross - a symbol of the Union of the Polarity

Star of David symbol from Uxmal, Yucatan, Mexico about A.D. 1000

Winged circle with feathered tail from Assyria about 800 B.C.

This find suggests that the knowledge of the union of the two earths was ancient knowledge that the Mayans either brought or later came from either immigrating or sea faring peoples from Assyria. Would you believe carvings have been found with some Mayans wearing ear spools—ear rings—with the star of David design? It's true. Late classic granite stela in Compache, Mexico dated 700 A.D. clearly show a star of David on ear spools covering nearly half the side of the model's face.

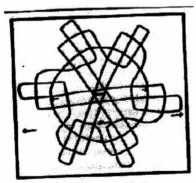

Continuously knit Mayan 6-pointed star at Piedras Negras showing the knitting together of the the spiritual and temporal worlds.

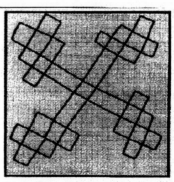

Mayan Union of the Polarity cross found at El Cayo. The seven squares at each end represent the seven aspects of the Creator.

The Mayan people had a clear understanding of the union of the polarity and represented it in nearly every form of art. In the above illustrations one can easily see the roots of Israel and the union of the spiritual and temporal earths.

In another excellent recovery of Mayan art, the dragon symbol was used as a representation of the union of the polarity. Two heads, very different in style and tone, were each used on one end of the dragon. As the classic glyph indicating the creation force, the dragon was used in Piedras Negras to show the artist's great care in keeping the people aware and reverent of this event.

Mayan two-headed dragon. Similar carving found on inscribed cliff at Piedras Negras. The dragon is the classic symbol of creation. The double headed dragon is a powerful symbol of the union of the polarity of two separate creations; one spiritual and one temporal.

In the ancient ruins of Monte Alban, Oaxaca, Mexico, a cross within a cross design was found. This same design appears on a Babylonian seal of the first dynasty dated at about 2000 B.C.

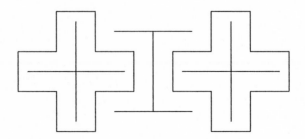

These cross symbols were pictorial testimony to the illiterate class of Mayans and other indigenous peoples of the union of the two planets. The Egyptian cross, carried by all the pharaohs, was the ankh, a symbol of the polarity union.

Egyptian Ankh Cross

In Peru the Andean Cross was the symbol of the polarity union.

Andean Cross

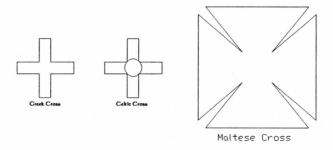

Greek Cross   Celtic Cross

Maltese Cross

The Greek cross, the Celtic cross and the Maltese cross, in all its variations, were symbols of the union of the two earths. Some of the huge stone pre-Christian Celtic crosses have carvings or figures of animals. One such cross was the Tuatha de Danann Cross of Clonmacnoise that was adorned with birds and other animals. Another was The North Cross, thirteen feet tall, with splendid figures of birds, deer, and a dog. These crosses are symbolically telling the story of the union of the two earths and the arrival of the new animals, from Noah's Ark, that joined with the surviving animals to replenish the temporal earth. The pillars and arches of Roslyn Chapel in Scotland are covered with hundreds of exquisitely carved figures of animals, leaves, plants, fruit, and other architectural devices that symbolize the union of the polarity and the arrival of the new animals. The ceiling is covered with multiple stars of David and the five-pointed star or pentacle. Some crosses combined circles and triangles to represent the union symbol and are practically universal throughout Asia, Europe, and America. The Buddhist wheel of life is composed of two crosses superimposed, a symbol of polarity union.

On the Mandalas of the Tibetans, heaven is laid out in the form of a cross, another ancient symbol of the union of the polarity.

India has preserved the cross in a great number of its temples, like the churches and cathedrals of Christendom, which are raised from cruciform foundations, all symbols of the ancient knowledge of polarity union.

In his book, *The Return of the Serpents of Wisdom,* Mark Pinkham states, "In the scriptures of Mesopotamia the cross upon which the dragon Enki coiled was the Gish Gana, the foliated tree which united Heaven and Earth."[21]

In fact, the cross, in its most basic and universal meaning, is a horizontal line that represents matter penetrated by a vertical line that represents spirit.

The ancients believed that a sexual union of the two earths took place as the spiritual earth was the male principle and the temporal earth was the female principle. World wide, the ancients further symbolized the sexual union of the two earths by carving and erecting phallic structures in stone, in the building of the tall Irish round towers, and in the erecting of Egyptian obelisks, all symbolic phallic structures denoting the union of the polarity. The female principle, or the temporal earth, was represented by a circle, or oval, called a yoni.

After the New World was discovered, the Knight's Templars having helped create an independent Scotland, then a "New Scotland", sought to create the "New Jerusalem" in the land of America under a new order called the Freemasons. America's founding fathers were almost all Freemasons. Freemasonry spread like "wild fire" among the colonists and would infiltrate into the whole of colonial administration, society and culture, which prevailed into the early part of the nineteenth century. This was aided by the immigration of British regulars into North America who brought with them Freemasonry of "higher degree" Freemasonry warranted by the Irish Grand Lodge.

Historians have often scratched their heads as to why the British lost the battle in the Revolutionary War and many reasons have been given in the history books except the right one. The colonists were out-ranked, often out-done in military tactics, poorly outfitted with military gear and provisions, and probably not as well trained. So why did the British lose the four-year war to this rag tag colonial army? The fact is they could have won if they really wanted to, but both sides were fellow Englishmen and Freemasons fighting brother against brother and fighting against principles they both believed; democracy and freedom. But the main reason the British Army did not seek victory for the crown is they knew the prophecy that America was destined to be the New Jerusalem, the latter-day fertile ground on which the restored gospel of Jesus Christ could bloom unfettered by government designs, and they didn't want to fight against God and be blamed for the failure of the prophecy in which prophecy all Freemasons believed.

We are not suggesting that the British commanders were guilty of treason but rather they gave only half-hearted approaches in a war that neither side wanted. They were professional soldiers prepared to carry out their government's commands and to do their duty, however reluctant and as narrow as possible.[22] The duty performed by British troops under royal command was marginal enough to ensure America's independence in fulfillment of the prophetic New World Order. The authors want to make perfectly clear that America owes her independence to Freemasonry and the prophecy.

It is a historical fact that numerous British officers ordered their firing lines to aim high into the trees above the heads of the revolutionary soldiers while at the same time being directly targeted by colonial long rifles. On one occasion, a colonial patrol outside Valley Forge stumbled across and captured a British patrol. The belongings of the prisoners were brought to Washington's headquarters, and were usually distributed among his poorly outfitted troops. Among the belongings of these British prisoners was a carrying bag with symbols of Freemasonry on the outside. Washington ordered the owner of the bag to be brought to him. After meeting nearly all afternoon and evening in brotherhood, the prisoner was returned to custody. After the war, this soldier not only commented about the excellent care he received as a prisoner of war, but at his deathbed requested that his friend and Freemason brother Washington administer to him. Washington did heed the request and dressed the British veteran in the appropriate clothing as per the ordinances of the order.

The first National Seal of the United States adopted the symbol of the Phoenix, the bird of freedom, whose ancient sacred number was 13.

The Phoenix bird was believed by the ancients to have emanated from the Solar Spirit or Life Force. It carried in its mouth a banner of thirteen letters, "E.Pluribus Unum," "out of many, one." In its talons were 13 arrows with 13 leaves on an olive branch with 13 fruits. There were 13 original colonies, 13 signers of the Declaration of Independence, 13 divided courses in the pyramid and 13 letters, "Annuit Ceoptis," meaning: "He (God) hath prospered our beginning." The number 13 also represented the 13 tribes (Joseph's sons, Ephraim and Manasseh received a double portion increasing the tribes to 13 in number) of Israel because our founding forefathers believed that America was destined to become the New Jerusalem and gathering place for the New Israel comprised of descendants of Abraham, Isaac and Jacob. (Israelites) A flag was designed of 13 stars and 13 alternating red and white stripes which represented the androgynous (united) nature of the Phoenix (red-male, white-female).

In 1841 the Phoenix was replaced by the symbol of the eagle.

The eagle has a shield of 13 bars. Above the head of the eagle is a set of 13 stars arranged in the form of a Star of David, an ancient symbol of the union of the polarity. The national seal is carried every day by almost all American's because it appears on the back of the one dollar bill currency.

The Freemasons incorporated the ancient knowledge of the union of the polarity into the layout of the city of Washington, using a circular design with streets radiating from the Capitol and from the President's home. These were designed to produce specifically octagonal patterns incorporating the particular cross used by Masonic Templars, which was a sign of the union of the polarity. The round rotunda (spirit) of the Capitol building with its long rectangular halls (matter) that adjoin on either side are also symbols of the union of the polarity. [23] The use of circles, rectangles, squares, octagons, and triangles, incorporated with crosses is often referred to as "sacred geometry". Sacred geometry also employs

All seeing eye with
13-step pyramid

a relationship called the *golden proportion*, a never-ending number of 1. 618…—also called *phi* in geometrical designs. These proportions are found in the natural design and structures of atoms to galaxies. The structure of the human body embodies this proportion. Since the *golden proportion* is found practically everywhere in the design of nature, it was considered sacred geometry and therefore was used in the design of ancient temples, pyramids, cathedrals, and cities.

At Washington's funeral he wore his Masonic apron that had among its symbolism the four pillars of Solomon's Temple—where the manuscripts of the *Divine Book of Wisdom* were found in their hollow insides—with a picture of Noah's Ark between them on one side and a star between the two other pillars on the other side. Two crossed swords were laid across the apron, symbolizing the union of the polarity.

The writings of Josephus of the building of Solomon's temple states more about the hollow pillars where the *Divine Book of Wisdom* was found by the Templars as follows, "Moreover, this Huram—a craftsman from Tyre—made two {hollow} pillars, whose outsides were of brass, and the thickness of the brass was four fingers' breath, and the height of the pillars was eighteen cubits and their circumference twelve cubits; but there was cast with each of their chapters lily work that stood upon the pillar, and it was elevated five cubits, around which there was network interwoven with small palms, made of brass, and covered the lily work. To this also were hung two hundred pomegranates, in two rows. The one of these pillars he set at the entrance of the porch on the oath, and called it *Jachin* and the other at the left hand, and called it *Boaz. "24* Josephus only wrote of two hollow columns at the entrance of the temple but there were two other hollow columns. These pillars are almost always depicted on Masonic ceremonial aprons, such as the one Washington wore.

The apron, Washington wore, symbolically associates Noah's Ark with the four pillars of Solomon's Temple, where the ancient manuscripts were found. The star, in this case a five pointed star or pentacle, represents the feminine earth or the female (Yin) principle of the union of the polarity. The Washington Monument, a 600 foot "frozen snake" Egyptian obelisk, which supports the spiraling "Life Force" between heaven and earth by acting as an antenna, was erected in his honor.

Noah's Ark

The Masonic Apron of George Washington

The book, *Genesis and the Chinese*, by C. H. Kang in his 1950 publication studied Chinese pictographic characters that give convincing evidence of the flood account of *Genesis* and the union of the polarity. The following are two examples within the Flood account.

Eight + United + Earth = Total.......... + Water = Flood

Vessel + Eight + Mouth = Boat

World wide the ancient knowledge of the "Life Force" and union of the polarity was symbolized in its structures that were built over dragon lairs where the spiraling "Life Force" was strongest. (readers are referred back to the chapter on *The Interaction of the Universes*) It was believed that the Creator or Life Force divided itself into two parts that opposed each other starting the concept of the Yin and Yang or female and male counterparts and other opposing principles such as day and night, hot and cold, life and death, good and evil etc. When these two opposing principles were in balance there was harmony and balance in all things. The Creator, symbolized by the serpent or snake motif, divided itself into two parts or twins that were symbolized differently in different cultures.

In the Mayan culture the Serpent Twins evolved into the Twin Boys, Xbalenque and Hunapu, of the *Popul Vuh*. To the Incas it was the two headed serpent.

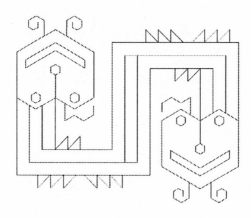

Incan Two-Headed Serpent

In Asia it was the two headed dragon, the Yin and Yang. In the Mediterranean region it was the Kaberoi and Dioskuri Twin Boys. The Aswin Twin Boys were the Hindu version. The Zunis of the southwestern United States call the twins the Ahayuta. The Iroquois call the Twins, Taweskare and Tsentsa. The Hopi's version of the Twin Boys were Poqanghoya and Polanghoya. In Greece they were known as Agatho Daimon and Agatho Tyche. In Mexico it was the two halves of Coatlicue. The Persians depicted them as twin serpents, Ahriman and Orzmund. The Egyptians depicted twin serpents emanating out of a winged solar disc. These are just a few of the names given to the twin children or twin aspects of the Creator or Life Force. Sometimes they were symbolized as twin spirals.

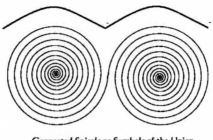

Connected Spirals as Symbols of the Union
of the Polarity

The ancients believed that without the checks and balances the opposing forces imposed upon each other, the creation would soon collapse.[25] Each principle was either male or female. When these two principles united it was called the union of the polarity. When the polarity is unified, or the two parts of the creator once again become one, the primal serpent or creator is said to be androgynous—both male and female. The Creator could therefore emanate itself as either male or female, as a God or Goddess.

Armed with the above knowledge we can interpret many of the ancient structures that symbolized the spiraling "Life Force" and polarity union. It is only here that we can find "rock hard" evidence of the ancient knowledge of the merging of the two earths in a corrupted form. Literally, there are hundreds of ancient sites world wide, so we will address only a few of the most well known and most powerful dragon lairs.

In Athens Greece, the Acropolis and its pillared Parthenon was built over a dragon's lair and oriented towards the Pleiades. The Parthenon was a home to the serpent "Life Force", whose pillars the spiraling "Life Force" moved up and down from the roof to the floor uniting heaven to the earth.

The island of Malta was originally called Ma Lato, the place of Mother Lato, where a serpent creator goddess was worshiped. The entire island is a dragon's lair and its many temples dedicated to the goddess are inscribed with recurring spirals upon their blocks which identify it as such. Single spirals represented the goddess creator and connecting pairs of spirals represented her dragon twins. A spiraling vortex is a unifying place of heaven and earth.[26]

The megaliths of Britain were all connected by ley or dragon lines. These include the sites of Stonehenge, Avebury, Bath, Glastonbury, and those on the sacred islands of Wright, Man, Anglesey, and Iona. We will explore only three, Stonehenge, Avebury, and Glastonbury Tor.

Stonehenge is very ancient. It was probably built about 2,600 to 3,000 B.C., maybe older. It was built over a very powerful dragon's lair symbolized by its circular construction that represented a yoni (female principle). The standing stones acted as a vortex facilitator to disperse the energy. So powerful is the vortex that the ancients considered it a portal or gate to heaven and a sacred place. Many have speculated as to the purpose of this structure. We propose this one. Stonehenge had a dual purpose. One we have already identified; a vortex for the spiraling "Life Force". Its second purpose was not just as an astrological site used to monitor the movement of the sun and moon to predict season change, but most of all to warn of an impending cataclysm. Failure of either the sun or moon to fall or be seen on certain markers at a given predictable time would indicate that once again the "earth was sinking to its destruction" and another flood was coming. The following is taken from the Discovery channel website, *Mystic Places*, that describes the movements of the Sun and Moon, but most noticeably, the Sun.

> "For most parts of the year, the sunrise can't even be seen from the centre of the monument. But on the longest day of the year, the June 21[27] summer solstice, the rising sun appears behind one of the main stones, creating the illusion that it is balancing on the stone.
>
> This stone, called the "Heel Stone", sits along a wide laneway, known as the Avenue, that extends from the northeast corner of the main monument. The rising Sun creeps up the length of the rock, creating a shadow that extends deep into the heart of five pairs of sarsen stone trilithons----two pillar stones with one laid across the top---in the shape of a horseshoe that opens up towards the rising sun.
>
> Just as the Sun clears the horizon, it appears to hover momentarily on the tip of the Heel Stone. A few days later, on midsummer's day, the sun will appear once again, but this time, it will begin to move to the right of the Heel Stone. The same phenomenon hap-

pens again during the winter solstice, only it's in the opposite direction and a sunset. But both indicate a change of season.

Aside from the sarsen horseshoe trilithons that open in the direction of the sunrise, there are four stones, called "Station Stones" that may have played an astronomical role. These were placed in a rectangle around the main monument, within the ditch and bank that surrounds the circle of stones. These are believed to point out the moonrise, moonset, sunrise and sunset. Only two stand today." (viewed by the authors Jan. 25, 2003)

Are these not markers monitoring the movements of the Sun and Moon? The rectangle of Station Stones would represent the male principle of the Life Force.

Avebury is built in the shape of a great slithering serpent that represents the "Life Force" or Creator. It was built as a marker to indicate that this was a powerful dragon's lair. On the back of the serpent is a large circle or disc that contains two smaller circles within. The large circle represents the female principle or yoni of the vortex. The smaller circles would represent the dual aspects of the Creator or "Twins". These all symbolize the union of the polarity. The structure was probably built around 2,600 B.C., which places its construction near to the time of the construction of Stonehenge with the possibility that the two sites had the same architects, although we feel that Stonehenge was a little older. An upcoming chapter, *The Giants,* will expand on the builders of these structures.

The Avebury Stone Serpent

Glastonbury Tor is a mound or tor so named because tors also symbolize the union of the polarity or male/female principles and were also vortex facilitators. This is because it was built over a dragon's lair in the shape of a truncated pyramid with seven tiers—seven curves or coils, the seven aspects of the creator—gracefully winding, in one continuous spiral like a serpents body, to its lofty peak. The focus of power was at its peak, the place where heaven (the spirit earth) and earth (physical earth) united and the "Life Force" was the strongest. It was also recognized to be an omphalos or a navel of the universe that symbolized the place where the spirit earth entered the temporal earth or was attached.

Legend has it that the Essene Joseph of Aramithea, under the instruction of the Apostle Philip, arrived here shortly after the crucifixion of Jesus Christ bringing with him the Holy Grail. When he arrived at the Glastonbury vortex he planted his staff into its Life Force charged soil and it immediately blossomed forth flowers, a sign from God that this was his place of destination. Joseph built the first Christian Church of Europe there. The Grail is reputed to be hidden within Chalice Well or Chalice Hill near Glastonbury. [28]

Another megalithic site worthy of mention is Callanish that is located on Lewis Island in the Outer Hebrides. The site seems to have been built in the form of a huge Celtic cross with a stone circle in the center where the cross intersects. The stone circle would indicate that the site was over a dragon's lair or vortex and was a yoni, the female principle. The cross (male principle) and the circle (female principle) would symbolize the union of the two earths or polarity union.

In Egypt the first Pharaoh was named Pharaoh and he was the first to start a long line of kings bearing the title of Pharaoh, which name means the son of a God. As said before in a previous chapter, the Egyptians viewed Noah and his sons as gods. The first Pharaoh was a great grandson of Noah, or son of a God, and was therefore like a God himself. All Pharaohs thereafter took on the personification of a God and claimed divine descent.

When a Pharaoh reigned for 30 years or more, some chose to renew their authority through the observance of the ritual of Sed. This ritual was performed in Memphis and required the Pharaoh to re-enact the union of upper and lower Egypt by sitting alternately on two thrones, thereby re-uniting spirit and matter which also symbolized the renewal of the "Life Force".[29]

This symbolism of the ritual, (upper Egypt meaning the spirit earth) and (lower Egypt meaning the temporal earth) came from the first Pharaoh, great grandson of Noah, who had direct knowledge of the event. This great grandson was a son of Egyptus, who was the daughter of Ham and Egyptus. This daughter, having being named after her mother, Egyptus, discovered upper Egypt, which was under water, and settled her sons in it. Her eldest son was named Pharaoh who established the first government and kingdom in the land. He was a righteous man who ruled wisely and justly all his days. (*Book of Abraham*) According to Herodotus, Pharaoh or King Menes was the first historical king of Egypt. He may have been the son of the first Pharaoh. He ruled over upper Egypt, and attributed to have drained the papyrus swamp land by damming up and changing the course of the Nile River. On the reclaimed land he then founded the city of Memphis. At about 3,000 B.C., he defeated the king of lower Egypt and united both lands of which he ruled. After he died his son ruled, and after his death his grandson ruled, etc. continuing the line through 30 dynasties. This line of Pharaohs believed that the royal family descended from the Gods and had royal blood. If Pharaoh Menes was the great, great grandson of Noah, then the ritual of Sed would have started in his reign, and the knowledge of the two earths merging would have been passed down to him from his father, Pharaoh, the great grandson of Ham. And as claimed, they were of noble birth or royal blood descent.

An Egyptian priest, named Manetho, wrote a history of Egypt in the third century B.C. He states that the Pharaohs were the direct heirs of gods who came to earth in remote millennia.

The Pharaohs wore a beard that was the body of a snake which ended with the tail feathers of a bird. Attached to their crowns was a serpent head (uraus). These two symbols represented the union of the polarity, feathers being the female principle and the male principle represented by the snake. They were feathered serpents or Gods. A symbol of their authority was the ankh, a version of the androgynous cross, which also symbolized polarity union.

The whole Giza Plateau in Egypt is a huge vortex. Because it is located in the exact geographical center of the land mass it sits upon, and because of its significant position upon the world grid, the Giza vortex is a perfect balance of north and south and a unifying point of the polarity.[30] It wasn't by accident that the Great Pyramid and Sphinx were built at that location. The great pyramid was actually constructed at the exact epicenter of the land masses of the earth. Without the perspective of extreme high-altitude, it is a wonder that such an ancient civilization could determine this location. The architects had the knowledge passed to them from the family of pharaohs who received it at the time of the union of the polarity. The Sphinx first marked the site as a vortex and later the Great Pyramid was built and constructed out of materials that amplified the "Life Force" energy. It became the perfect site to house the library known as the "Hall of Records", which library has yet to be discovered but known among the ancients to exist in an under ground chamber beneath the Sphinx.

The Intiwatana or the "Hitching Post of the Sun", at Machu Picchu, Peru, marks one of the most powerful vortex sites in the world. The vibrating stone is carved in the form of a serpent and was once surmounted with a round sun disc. It is said that this site marks a portal or interdimensional gate to heaven. Ancient shaman performed rituals in vortexes that opened portals into the spiritual dimensions. Once a portal was opened, it was reused generation after generation. It was believed a portal would unite the material and spiritual worlds or polarity union.

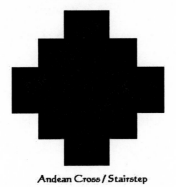

Andean Cross / Stairstep

Cuzco Peru, had an identical copy of the Intiwatana but the Spaniards destroyed the rock. Cuzco was built on a powerful vortex by the Incas. The city was sacred because it also marked an interdimensional gate to heaven and was believed to be the "navel of the world" (Cuzco means navel) or the place where the spirit earth entered the temporal earth and was attached. This ancient knowledge has long been lost and remembered today only as a "navel". The Incan symbol of the union of the polarity was the two headed serpent.

Tiahuanako, Bolivia, is also situated on a very powerful vortex at the convergence of multiple dragon lines. In the center of this vortex a huge 150 foot high pyramid was built in seven tiers and in the shape of a colossal tau cross with cross/stairstep pattern, symbols of the union of the polarity.[31]

The Mayan ruins of Lamanai, Belize, has a stone seat, located at the western side of a pyramid in the city, that is carved with the symbols of the Yin and Yang, symbols of polarity union. The site dates back to 100 B.C. One of your authors has sat upon this seat twice, having visited the site at two different times on expeditions.

Although there were Mayan temples dedicated to the primal dragon, Life Force, or creator, there were also jaguar temples. It is easy to identify the jaguar temples because the jaguar carvings usually had cat paws, ears, and tail. Not so with the dragon masks. There is a Mayan temple (pyramid K-5) in Piedras Negras, Guatemala, that was covered with giant stucco masks. That region of the country has many large caves, which are associated in ancient traditions as the lairs of dragons. These lairs are vortexes where the spiraling Life Force energy is strong. It is the opinion of the authors that archeologists have erred in calling these giant stucco masks, jaguar masks. They look much different, like a dragon.

In the classic Mayan ruins of Piedras Negras (400 B.C. – 1,400 A.D.), located on the Usumacinta River in Guatemala, (ancient name Y-Okib which means "large entrance" or cave) is a temple pyramid identified as K-5 that has two surviving giant stucco masks adorning its walls. One mask is in good shape, considering its age; the other only half there. Anciently there were probably many more on each of the pyramid's multiple terraces. Similar masks were discovered in the Mayan ruins at Cerros, Belize, on a late preclassic temple circa 50 B.C., Uaxactun, and in other Mayan sites in the Northern Yucatan and the Mexican Highlands. Archeologists have identified these giant stucco masks as jaguar masks, but your authors disagree. We believe they are clearly dragon masks; see picture below.

Giant stucco mask on corner of Pyramid K-5 mistakenly interpreted as a Jaguar. The actual depiction is that of a dragon, the ancient symbol of the Creator God or the Creative Life Force.

Olmec dragon mask showing feathered serpent symbolizing the Life Force or Creator God. Originating circa 2700 B.C. it may have left by the first settlers from the Orient.

275

Nearby, are spiral circle carvings that identify this site as a vortex or dragon's lair where the spiraling Life Force is strong. Most all temples are located in ceremonial centers near vortexes that the Mayans and other cultures believed to be gates or portals into the spiritual dimensions. It appears that the worship of the creator, or Life Force, symbolized by the dragon, or serpent, was practiced world-wide anciently. In addition, most of the ancient world knew about the union of the polarity, however, there was not much evidence that they understood it very well.

Most of Piedras Negras is unexplored and covered with dense jungle. The only way into the site is by a nine-hour, hair-raising white water river boat trip on weekly arranged expeditions that requires camping on the river banks in primitive conditions. Tourists visit only the small accessible part that has been excavated and restored. The entire site contains vast unexplored cave systems, a large cenote, and possible tunnels or caves into a mountain whose entrance is covered by a pyramid. This pyramid is built over an older pyramid that appears to cover a blocked entrance to either a tunnel or a cave entrance. It is located in a remote unexplored section of the site, just recently discovered. Perhaps is this tunnel or cave from which Y-Okib derives its name. Your authors believe that this particular pyramid sire is a good candidate for the location of the Yucatan Hall of Records.

The Mayan ruins of Piedras Negras has a carving on Inscribed Cliff of a two headed dragon. (See two-headed dragon illustration) In the Mayan ruins of Palenque there are two murals showing Pacal seated on a two headed jaguar bench and a two headed jaguar bench (below), all symbols of the union of the polarity. The two heads represent the "Twins" or dual aspects of the Creator or "Life Force."

The coronation of Pacal

Similarly, the Hindu and Chinese depict two-headed benches or seats and two-headed dragons in many of their works of art.

Teotihuacán, Mexico was laid out in the form of a huge cross. The Pyramid of the Sun was built near the center of the site over a powerful vortex that symbolized polarity union.

Tenochtitlan, the capital city of the Aztec empire, was also built on a vortex. The principal pyramid, Temple of Tenochtitlan, was built over it. At the summit of the Templo Mayor temple, it divided into two separate temples symbolic of the Serpent Twins, Huitzilopochtli (male principle) and Tlaloc (female principle), symbols of polarity unity.[32]

Quetzalcoatl was known as the "Plumed Serpent" and his symbol was the Tau cross. The feathers represented the temporal principle and the serpent represented the spiritual principle. These three principles, feathers, serpent, and Tau cross were all symbols of polarity union. A "Plumed Serpent" is the title given only to creator gods.

The feathered serpernt is a universal symbol for the Creator God, and is widely used in descriptions of the union of the polarity. The joining of the spiritual earth with the physical earth is infused into so many ancient cultures, there is little doubt of common origin--the beginning of time.

Incan Feathered Serpent

Egyptian Feathered Serpent

Mississippi Mound-Builder's Feathered Serpent

Mayan art on temple wall friezes symbolized the union of the polarity. Below are some illustrations.

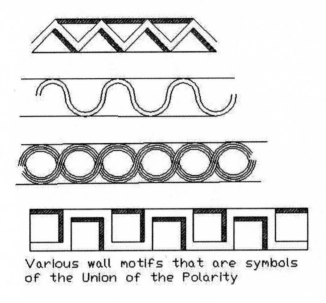

Various wall motifs that are symbols of the Union of the Polarity

The largest pyramid on Earth is near Xian, China. It is more than twice the size of the Great Pyramid of Giza. Although its lower chamber has never been opened, there are ample descriptions of what is contained therein. The emperor entombed inside is provided a scale model of his city. The ceiling of the tomb is said to be a layout of the constellations of Orion and other significant formations in jewels embedded in the stone. The buildings, roads, and shrines of the city are all reconstructed and laid out on the floor of the chamber, stretching for hundreds of yards under the pyramid. The rivers and streams are excavated into the stone floor and wind throughout the city exactly as they did during his reign. They are filled with mercury. Along the walls of the chamber are large oil-burning lamps that are fueled from large reservoirs of whale oil. The original Chinese designers wanted the light from the lamps to reflect off the rivers of mercury and laminate the jewels on the ceiling and make the city come alive forever. The Chinese government forbids tourists from visiting the pyramid. No archeological work has ever been done to substantiate any of the legends of its construction.

In fact, thousands of stone circles, megalithic temples and pyramids dot Asia and are seldom seen by tourist. Pyramid complexes were built as far north as Siberia, Russia. India at one time had a pyramid complex named Bindh Madhu that sat perhaps thousands of years overlooking the Ganges before the fanatical Moslem emperor Aurangzed destroyed them. In Cambodia another pyramid complex was built, now known as the ruins of Angor Wat. [33] The ancient knowledge of the two earths merging was apparently understood by the children of the East, descendants of Japheth, Ham, and Abraham's children by his concubines.

Recently in Florida, on the Miami River bank, where the river joins Biscayne Bay near downtown Miami, the remains of a stone circle were found. Yoni's are always built on a vortex and symbolize polarity union. It is the first stone circle known to exist in the eastern part of North America and at present time, builders unknown.

The word Yoga means united. The disciplines of Yoga awaken the Kundalini ("Life Force" or Dragon) from its seat at the base of the spine and kindles the serpent fire. Through meditation the serpent fire can easily move up the spine, through seven major chakras or vortexes, in spirals like the serpents of the caduceus, and when they unite at the top of the head, balance or polarity union is produced, and spiritual enlightenment can be attained by "Life Force" stimulation of the inner or third eye, sometimes referred to as the spiritual eye, located in the forehead. Biblical scripture supports this concept in the following verse, "If thine eye be single thy whole body is filled with light." [34]Hindu's mark the third eye by the black dot, called the tika, on their forehead and acknowledge it as the "seat of wisdom." Balance promotes health and it is the same principle that the art of acupuncture uses to restore the "Life Force" balance within the human body through uniting the energy channels to produce polarity union. It is restoring the Yin and Yang. It is an ancient meditative art and science. Does it work? Millions of people seem to think so.

The caduceus is also an ancient symbol of the union of the two earths. It has the "Twin" serpents spiraling up the rod of Mercury representing the union of the two dual aspects of the Serpent Creator or "Life Force", promoting balance. The feathered wings at the top of the staff represent the female principle or matter and the serpent represents the male principle or spirit. It is a feathered serpent, ancient symbol of polarity union. Today it is the symbol of the medical profession.

It is said that when St. Patrick arrived in Ireland, he killed all the snakes. He literally didn't kill live snakes but he killed the old serpent and nature worship religions through the teachings of Christianity. Christianity simply replaced Druidism and the old religions. In England, it is said that St. Michael slew the dragon. He didn't kill a dragon literally, but again he symbolically killed the old dragon religion through the new teachings of Christianity. When Christian churches were built on ley or dragon lines, as St. Michael's Church in England, they symbolically slew the Dragon religion. In the process of replacing the old religions with Christianity, the remnants of the ancient knowledge of the merging of the two earths were inadvertently lost in Christian countries.

Fortunately, *The Zohar* or *Book of Splendor*, preserved the event, though poorly understood, in many of its writings. We owe a debt of gratitude to the ancients for the symbolism carved into stone and for the preservation of ancient records that helped us "unveil" the great mystery. Five years ago it would have been impossible to read many of these records as they had not been available in English.

The irony is that we have had the ancient knowledge all along through symbolism, even around us daily by the exchange of the one dollar bill American currency. Almost world wide the symbolism was chiseled in stone as silent testimonials and in the sacred geometry of the Freemasons for all to "see." As the result of the union of the polarity, one might say that we are now living in two different worlds at the same time.

Those spiritually awake have already "seen" through the window of wisdom with their inner or third eye. For the rest who are spiritually asleep, it is as the old saying goes... "They can't seem to see the forest for all the trees."

"Earth with its Spiritual Shadow"

End Notes

[1] *(Psalms 104.5)*

[2] *(Sanhedrin 65b) (Excerpted from Sefer Yetzirah by Aryeh Kaplan ©1997 with permission of Red Wheel/Weiser, York Beach, ME and Boston, MA, pp.42, 60)*

[3] *(The Zohar* et al Vol. 1V, p. 253)

[4] *(The Zohar* et al Vol. 1V, p. 273)

[5] *(The Zohar* et al Vol. 1V, p. 287)

[6] *(The Zohar* et al Vol. 1V, p. 299) Parenthesis added.

[7] *(The Zohar* et al Vol. 111, p. 359)

[8] *(The Zohar* et al 1, p. XV)

[9] *(The Zohar* et al 1V, p. 19)

[10] *(The Zohar* et al Vol. 111, p. 383)

[11] *(Sefer Yetzirah,* et al p. 43)

[12] Excerpts taken from *The Return of The Serpents of Wisdom* ©1997 by Mark Amaru Pinkham. Published by Adventures Unlimited Press, Kempton, ILL. All rights reserved. Used by permission of the publisher, p. 310)

[13] *Lost Cities & Ancient Mysteries of South America,* ©1986 by David Hatcher Childress. Published by Adventures Unlimited Press, Kempton, ILL. All rights reserved. p. 356)

[14] *(The Return of the Serpents of Wisdom,* et al p. 228, as cited in *The New View Over Atlantis*, John Mitchell, 1983, Harper and Row, San Francisco, Ca.)

[15] *(The Return of the Serpents of Wisdom,* et al pp., 227, 228, as cited in The Holy *Bible,* Cambridge University Press, NYC)

[16] *(The Return of the Serpents of Wisdom,* et al p. 268)

[17] *Lost Cities & Ancient Mysteries of South America,* ©1986 by David Hatcher Childress. Published by Adventures Unlimited Press, Kempton, ILL. All rights reserved. p. 406)

[18] *(The Temple and the Lodge,* Baigent and Leigh, Arcade Publishing, N.Y., 1989, pp.55, 56) *Lost Cities of Atlantis, Ancient Europe & The Mediterranean,* ©1996 by David Hatcher Childress. Published by Adventures Unlimited Press, Kempton, ILL. . , p. 258)

[19] *Lost Cities & Ancient Mysteries of South America,* ©1986 by David Hatcher Childress. Published by Adventures Unlimited Press, Kempton, ILL. All rights reserved., p. 398)

[20] Excerpts taken from *The Return of The Serpents of Wisdom* ©1997 by Mark Amaru Pinkham. Published by Adventures Unlimited Press, Kempton, ILL. All rights reserved. Used by permission of the publisher, p. 350)

[21] *(The Return of the Serpents of Wisdom,* et al p. 67)

[22] *(New Evidences of Christ in Ancient America,* Yorgason, Warren, & Brown, *Book of Mormon* Research Foundation, 1999, pp.204-205)

[23] (et al p. 325)

[24] *(The Temple and The Lodge,* Michael Baigent and Richard Leigh, Arcade Publishing, N.Y., 1989, p. 216)

[25] (*The Return of the Serpents of Wisdom*, et al p. 296) (*The Temple and The Lodge*, p. 262)

[26] (*The Complete Writings of Josephus,* translated by William Whiston with commentary by Paul L. Maier, Kregel Publications, 1000, p. 273)

[27] (*The Return of the Serpents of Wisdom*, et al pp.341, 342, 343, 344, 345)

[28] (*The Return of the Serpents of Wisdom*, et al pp.78, 79)

[29] (*The Serpents of Wisdom*, et al pp.78, 70, 80, 81, 83)

[30] (*The Return of the Serpents of Wisdom*, et al p. 99)

[31] (*The Return of the Serpents of Wisdom*, et al p. 75)

[32] (*The Return of the Serpents of Wisdom*, et al pp.88, 89)

[33] (*The Return of the Serpents of Wisdom*, et al pp.87, 88)

[34] (*The Return of the Serpents of Wisdom*, et al pp.94, 95)

[35] KJV *(Matthew 6:22)*

The Books

When Noah exited the ark and established a more permanent residence in the area, he brought all the books or sacred manuscripts with him; books that had been entrusted to him for safe keeping from his forefathers, beginning with Adam and were passed down ten generations into his care. Exactly how many books, scrolls, or manuscripts were entrusted into his care is not known but presumably he had enough copies to give each son, as the Lord had commanded much earlier, and to have at least one copy of every manuscript for himself.

Do these manuscripts or writings exist today or have they disappeared through the ravages of time? If they were important enough for the Lord to assign guardian angels to protect them through the deluge, would they not still be protected in our time? Just where are these writings, and what messages or information do they contain? Surely such important manuscripts must deliver powerful wisdom and extremely important information worthy of

287

continued preservation, after all, some of this wisdom revealed the workings of God in creation; works and teachings that were not revealed even to the holy angels.

So far we have traced the history of these writings, some which were given to Adam from God, and many that Enoch wrote, and were passed down to Noah. Now let us trace them forward from Noah to present time. When the evidence is examined through sacred and ancient writings, the paper trail becomes in some instances quiet clear. Other times the evidence is rather "murky" but supportive enough to form well educated guesses. It will be impossible to locate all the books or manuscripts but certainly some can be found and in the most surprising places! The question is can you accept the messages on these most ancient manuscripts even if they disagree with your own religious philosophy? Apparently this was a problem with the early Christian churches whose religious leaders intentionally suppressed some of the writings and persecuted those religious sects whose writings differed from their views. Do we not encounter these same prejudges today if any new views are expressed in religious theory that may be in conflict with old doctrine?

Enough said! Now let us pursue the whereabouts of these ancient books. Keep in mind that a *book* may not literally mean a book by today's standards. It may simply be a one page manuscript. It is doubtful if any of the original texts could have survived through several thousand years, unless it was written on stone tablets or medal plates. The ancient scribes, being aware of texts succumbing to age, copied the writings in successive generations to preserve them. In time some religious teachers added their own interpretations and embellishments in counterfeit copies of the original writings. The authors will expose known corrupted texts.

According to the 1840 translation of *The Book of Jasher*, from the Hebrew into English, Abram (Abraham) was sent to live with Noah and Seth from early childhood and he lived in their home for thirty-nine years. His father, Terrah, sent him there for his protection from King Nimrod, who sought his life, because his birth was announced by a new large star that the sages interpreted as the birth of a powerful leader whose seed would possess all the earth and perceived by Nimrod to be a threat to his kingdom.

While Abram remained with Noah and Seth, no one knew his whereabouts and in time King Nimrod forgot about the child Abram because Terrah gave the King an infant of his handmaid, which Nimrod slew, thinking that it was the infant Abram.[1]

During the thirty-nine years of living with Noah and Seth, Abram was taught the ways of the Lord and instructed in wisdom from the books that Noah had preserved. From these books he learned about the creation of the universes, the planets (other worlds), stars and moon. He learned their movements as to signs for seasons and their effect on the earth and each other (gravity). He learned of the creation of Adam and of the fallen nature of men. He studied the creative powers that enabled God to create. He received instruction in mathematics. He understood the spiritual creation and how Noah had arrived to this planet. Collectively, Noah, Seth, and Abram were the wisest and most righteous men of their time. Sadly, in a few generations, all men forsook the Lord and practiced idolatry except Noah, his household, and all those who were under his counsel. In time, after Noah died, Abraham sent his son Isaac to the household of Shem and Eber, the great grandson of Shem, to learn the ways of the Lord and his instructions. Abraham and Isaac remained there three years. Later Isaac sent his younger son, Jacob, to Shem and Eber for the same learning and counsel. Jacob remained in the house of Shem and Eber for thirty-two years. Esau, Jacob's brother, did not go, for he was not willing to go, and he remained in his father's house in the land of Canaan.[2]

Abraham probably obtained copies of Noah's books and studied them often. He was considered a master of the mysteries among the ancients. There are two ancient books, thought to be as old as the *Torah*, or the first five books of *The Holy Bible*, that are attributed to have been written by Abraham and exist today. This however, does not mean the entire text as we have it today was written by Abraham because various Rabbis down through the ages added their own written and oral interpretations as to the meaning of the various writings.

The first book is titled *SEFER YETZIRAH, The Book of Creation*. This book is without question the oldest and most mysterious of all Kabbalistic texts. A passage in this book clearly suggests that Abraham actually made use of the methods found in this text to create and is supported by the last stanza: "When Abraham...looked and probed...he was successful in creation..."

A Talmudic teaching that "Abraham had a great astrology in his heart, and all the kings of the east and west arose early at his door" further links him to this book as the *Sefer Yetzirah* is one of the primary ancient astrological sources and probably incorporates Abraham's astrological teachings. For centuries the *Sefer Yetzirah* was taught from memory as an oral teaching, and eventually was written in book form. The text was written word for word precisely as was taught orally by the wording of Rabbi Akiba.

The second book is now a set of five volumes called the *Sefer ha-zohar* or *THE ZOHAR, The Book of Splendor*. These compiled ancient writings are most significant because they present *The Zohar's* view of the correct method of interpretation of biblical passages, in which it says: "Woe unto those who see in the Law nothing but simple narratives and ordinary words...Every word of the Law contains an elevated sense and a sublime mystery..."

The back covers of all *The Zohar* books state the following: "Since its compilation, few works have exercised such an influence on people as *The Zohar*. It was said, if a mortal had written it, he must have been inspired from above."

Another ancient record, written by Abraham, was translated from a papyrus text taken from the catacombs of Egypt. It is called *The Book of Abraham* and contains priceless information about the nature of Deity, the creation, the pre-existence, and priesthood information not found in any other revelation now extant. The book is part of a collection called *THE PEARL OF GREAT PRICE*, scriptures used in the Church of Jesus Christ of Latter Day Saints.[3]

The Divine Book of Wisdom also contained knowledge in the practice of magic, alchemy, and divination. Abraham was fully aware of the dangers that could occur with the misuse and application of these mysteries into idolatrous uses. *The Talmud* states that Abraham wrote a tract dealing with idolatry that consisted of 400 chapters.[4]

In *Genesis* 25:6 it states "to the sons of the concubines that Abraham had, Abraham gave gifts, and he sent them away…to the lands of the east." These gifts consisted of occult mysteries which then quickly spread into eastern Asia, no doubt influencing the writings in the mystical Vedic scriptures of India.

Abraham was born in Mesopotamia and he also lived in Egypt. He had the reputation of being the greatest astrologer and mystic of his time and therefore it is natural to assume that he was familiar with all the mysteries of ancient Egypt. No doubt some of the occult knowledge, or possibly copies of the entire text of *The Divine Book of Wisdom* and other books had been passed down through a grandson of Ham to the magicians and astrologers of ancient Egypt. *The Book of Abraham* states that the land of Egypt was first discovered by a woman, who was the daughter of Ham, who was named Egyptus, which in the Chaldean signifies Egypt, which signifies that which is forbidden. When this woman discovered the land it was under water, who afterward settled her sons in it; and thus, from Ham, sprang that race which preserved the curse in the land. Now the first government of Egypt was established by Pharaoh—yes, his name was *Pharaoh*-- the eldest son of Egyptus, the daughter of Ham, and it was after the manner of the government of Ham, which was patriarchal. [5] These ancient Egyptian mystics and astrologers, who served in the king's palaces, jealously guarded their secrets. King Nimrod, son of Cush, grandson of Ham, had wise men and astrologers in his court, who presumably studied and gained their knowledge from the books being passed down through Ham, or even directly from Abraham as he spent time in Egypt acting as an advisor and teacher in matters pertaining to astrology.

Jacob, son of Isaac, gave his youngest son, Joseph, Ruler of Egypt, a treasured gift that had been passed down to him. It was the Rod of God. *The Book of Jasher* states: "From Noah this stick was given to Shem and his descendants, until it came into the hand of Abraham the Hebrew.

And when Abraham had given all he had to his son Isaac, he also gave to him this stick…

And when he (Jacob) went down to Egypt he took it into his hand and gave it to Joseph."[6]

Presumably Jacob, having studied under the counsel of Shem and Eber, also gave Joseph copies of the books passed down to him which greatly increased his knowledge in all things.

The Book of Jasher continues:

"And after the death of Joseph, the nobles of Egypt came into the house of Joseph, and the stick came into the hand of Reuel (Jethro), the Midianite, and when he went out of Egypt, he took it in his hand and planted it in his garden.

And all the mighty men of the Kinites tried to pluck it when they endeavored to get Zipporah his daughter, but they were unsuccessful.

So that stick remained planted in the garden of Reuel, until he came who had a right to it and took it.

And when Reuel saw the stick in the hand of Moses, he wondered at it, and he gave him his daughter Zipporah for a wife."[7]

Now to Moses, the Hebrew Prince of Egypt, who was raised in the palace of Pharaoh as a son. This Pharaoh was the grandson of the Pharaoh who made Joseph, son of Jacob, Ruler of Egypt.[8] Abraham was the ancestor of Moses, seven generations back. The father of Moses was Amram, who was the son of Kohath, whose father Levi was the son of Jacob, who was the son of Isaac, who was the son of Abraham.[9] It was apparent, at an early age, that Moses was an exceptional child. He was far brighter and advanced in wisdom for his age. His beauty of countenance made passers by turn around and stand still a great while to look on him. It was because of his beauty and wisdom that Pharaoh's daughter, having no children of her own, desired to adopt him for her son and make him heir to Pharaoh's kingdom.[10]

Moses was educated in the best schools and universities that Egypt had to offer. He was taught the sciences, mathematics, astrology, astronomy, alchemy, and was familiar with the magical occult sciences practiced by Balaam, the son of Beor, the magician. *The Book of Jasher* states that Balaam, the magician, his two sons, Jannes and Jambres, and eight brothers were the sorcerers and magicians who are mentioned in the *Book of the Law*, standing against Moses when the Lord brought the plagues upon Egypt.[11] Some of Moses's knowledge taught to him probably came from copies of the books that were passed down through the descendants of Ham and made their way into Egyptian libraries that were open only to the privileged few. These libraries were tended by sacred scribes of both nations, Hebrew and Egyptian.[12] Balaam most certainly learned the occult science and magic from the study of the ancient texts.

So favored was Moses by Pharaoh that he was appointed General of the Army against the Ethiopians, which he defeated in battle. Tharbis was the daughter of the King of the Ethiopians. She had seen him passing by and admired his courage and skills as a general and fell in love with him. The Egyptian army, having defeated the Ethiopians in battle, did not slacken but pursued the defeated Ethiopians to the point that they were in danger to being reduced to slavery and all sorts of destruction. Tharbis sent the most faithful of all her servants to Moses with a long entreaty and proposal of marriage. Moses thereupon accepted the offer, on condition she would procure the delivering up of the city; and gave her the assurance of an oath to take her to be his wife; and that when he had once taken possession of the city, he would not break his oath to her. No sooner was the agreement made, but it took effect immediately and when Moses had cut off the Ethiopians, he gave thanks to God, consummated his marriage, and led the Egyptians back to their own land.[13]

The fame of Moses had spread far and wide. When Moses reached the age of eighteen, he fled Egypt and escaped to the camp of Kilianus, King of Cush, which at that time was besieging Cush, because the people of Cush were led to rebel against the king by Balaam, the magician. *The Book of Jasher* records: "And the king and princes and all the fighting men loved Moses, for he was great and worthy, his stature was like a noble lion, his face was like the sun, and his strength was like that of a lion, and he was counselor to the king.

And at the end of nine years, Kilianus was seized with a mortal disease, and his illness prevailed over him, and he died on the seventh day.

And they wished to choose on that day a man for king from the army of Kilianus, and they found no object of their choice like Moses to reign over them.

Moses was twenty-seven years old when he began to reign over Cush, and forty years did he reign."[14]

Before God sent Moses to the Israelites, he was not only learned in all the wisdom of the Egyptians, but was also mighty in leadership, words and in deeds.[15]

The Book of Jasher continues:

"When Aaron and Moses first went to Egypt for an appearance before Pharaoh, Moses took the Rod of God with him and when they approached the king's house they found it was guarded by two lions chained at the door. Moses lifted up the Rod of God upon the lions and they were loosed and Moses and Aaron entered the king's house. The lions also came with them in joy. This astonished Pharaoh and he sent for Balaam, the magician, his sons and his brothers to seek their counsel in the matter. Balaam asked the king how they got past the lions, the king said," because they lifted up their rod against the lions and loosed them, and came to me, and the lions also rejoiced at them as a dog rejoices to meet his master." Balaam answered the king, saying, "These are none else than magicians like ourselves. Now therefore send for them, and let them come and we will try them.

The next morning Pharaoh sent for Moses and Aaron to come before the king, and they took the Rod of God, and came to the king and spoke to him, saying, "Thus said the Lord God of the Hebrews," "Send my people that they may serve me." Pharaoh demanded a sign to prove that they were messengers from God so that he might believe their words. Aaron then threw the rod out of his hand before Pharaoh and his magicians, and the rod turned into a serpent. The sorcerers saw this and they cast each man his rod upon the ground and they became serpents. Then the serpent of Aaron's rod lifted up its head and opened its mouth to swallow the rods of the magicians when Balaam said, "This thing has been from the days of old, that a serpent should swallow its fellow, and that living things devour each other. Now therefore restore it to a rod as it was at first, and we will also restore our rods as they were at first, and if thy rod shall swallow our rods, then shall we know that the spirit of God is in thee, and if not, thou art only an artificer like unto ourselves.

Aaron hastened and stretched forth his hand and caught hold of the serpent's tail and it became a rod in his hand, and the sorcerers did the like with their rods, and they got hold, each man of the tail of his serpent, and they became rods as at first. When they were restored to rods, the rod of Aaron swallowed up their rods."[16]

Although the collection of manuscripts commonly called *The Bible*, in several verses, refers to this staff as the rod of Aaron, it is called that because it was in the hands of Aaron who, at that time, was performing the works of the plagues by the use of this staff. Other verses have the rod in Moses' hand performing some of the plagues. Were it not for the *Book of Jasher* relating that Moses and Aaron took the Rod of God and visited Pharaoh, readers would think that there were two sticks or rods involved in the plagues. The writings of Josephus state that Moses threw down his rod on the ground in front of Pharaoh and the rod turned into a serpent and then devoured the other rods of the magicians.[17] It is not the intent of the authors to argue biblical doctrines but merely point out that the Rod of God and the rod of Aaron were one and the same. The plague of turning the waters into blood, the plague of frogs, and the plague of lice where all duplicated by Pharaoh's magicians.

The authors have retold this story to emphasize how powerful the magicians were in matching the actions of God. This was because they had studied occult magic from God's own book, *The Divine Book of Wisdom.* For centuries after there were magicians and sorcerers in every palace. Has any of this magical know how survived today? There is one book that seems to contain much of this knowledge and is considered the magician's handbook. It is titled, *THE GOLDEN DAWN*, as revealed by Israel Regardie. According to history and "myth", Dr. William Westcott discovered a cipher manuscript that led to the founding of the Hermetic Order of The Golden Dawn in 1888. Since its first publication, in 1937-40, this book has been the foundation for much of the magical and occult revival in the last half of the past century. Various magical groups, Wicca, the New Paganism, and serious students of the magical arts have studied its resource materials. Much of the material contained in the book appears to be ancient knowledge incorporating kabballah and magical working based on the Enochian Tablets. The introductory page states, "This book is not about Magic, it is Magic."

E. J. Clark & B. Alexander Agnew, PhD

In the British Museum, a set of ancient manuscripts were found that were translated by S. Liddell MacGregor Mathers, who was head of the Order of the Golden Dawn and who was a well known 19th century magician. These texts are believed to be the actual words and instructions of King Solomon himself. It is one of the most famous of all magical textbooks titled, *THE KEY OF SOLOMON, THE KING*. The back cover of the book states that he (Solomon) instructs his followers on how to summon and master spiritual powers, including how to obtain answers to problems from the spirit world. The book was first published in 1914.

These two books are the most famous in the magical circles and appear to incorporate ancient knowledge taken from Kabbalah. The practice of Kabbalah in itself is very ancient, the first written text was the *SEFER YETZIRAH*.

In 2000, Steve Savedow edited and translated into English from the ancient Hebrew in the rare 1701 Amsterdam edition of the book *Sepher Rezial* or *The Book of the Angel Rezial,* sometimes referred to as *The Divine Book of Wisdom* or *Adam's Book.* The book is a compilation of five books: *"The Book of the Vestment," "The Book of the Great Rezial," "The Holy Names," "The Book of the Mysteries,"* and *"The Book of the Signs of the Zodiac."* The back cover of the book states it is, "A diverse compendium of ancient Hebrew magical lore, this book was quiet possibly the original source for later traditional literature on angelic hierarchy, astrology, Qabalah, and Gematria." But, is it the original *Divine Book of Wisdom* or a forgery? Steve Savedow states in his introduction,

300

"Each of the five manuscripts within the 1701 edition was produced during a different time period, while the manner of writing Hebrew was changing and improving. In each manuscript there is a different form and dialect of biblical Hebrew. Some of the dialects are of a very ancient form that was never intended to be translated into English. The first sections appear to have been derived from the most modern manuscripts. The style of writing in each book is different, with the use of different prefixes and suffixes, and different verbs and word spellings, and the expressions varied from section to section. The later sections are a bit "degraded" in places, as they are transcriptions of the most ancient manuscripts, and they were the most difficult to translate. For these reasons, the translation may seem rough and the "flow" of the text changes throughout this book. However, there is no way to accurately interject terms for the sake of literary integrity without possibly making an incorrect assumption."

The original editor of this book states in the foreword, "This work is a compilation of several manuscripts that were passed down and corrected generation after generation." Your authors take note of the fact that some of the texts have admittedly been altered, no matter now ancient, which throws obvious doubt into their integrity, however part of the older text appears to be in tact. The bottom line, it is difficult to pass judgment on the older text. The readers will have to form their own opinion.

Many scholars and theologians have long debated who was the author of the book of *Genesis*, known as the first book of *The Torah*. It is generally accepted in these circles that Moses was attributed to be the author, but was he? There is evidence that indicates he was not the author of some of the chapters, but instead was a caretaker of those manuscripts.

The opening remarks of chapter 5 of *Genesis* states; "This is the book of the generations of Adam." It is a book consisting of about one page of manuscript and THIS IS THE BOOK written by Adam. It was one of the books passed down from Adam to Noah before the deluge.[18] Moses saw to it that this manuscript or book would be included in the creation accounts.

In light of this evidence, it appears that Moses had copies of the books and manuscripts that had been passed down. The first 2 chapters of *Genesis* are the creation accounts and these accounts were probably taken from *The Divine Book of Wisdom,* written by God. Chapter 3 of *Genesis* appears to have been written by Adam and may have been taken from the *Book of Adam's Remembrance* or it may be the book in its entirety.

Another latter day record, written by Moses, called *The Book of Moses*, is part of the collection known as *The Pearl of Great Price*, scriptures used by the Church of Jesus Christ of Latter Day Saints. In this record Moses covers the same events recorded in the first six chapters of *Genesis* but the account revealed in latter-days has been so enlarged and contains so much new material that it radically changes the whole perspective of the Lord's dealing with Adam and the early patriarchs.[19]

In *The Book of Moses* chapter 6:8-9, Moses states that the book of the generations of Adam was kept of the children of God and was written by Adam being inspired by the Holy Ghost. This clearly states that he was not the author of chapter 5 of *Genesis*.

Why do the writings of Moses in the latter-day account contain so much new material than found in the *Old Testament* accounts of the first 2 chapters of *Genesis*? Perhaps Moses also questioned the brevity of the creation accounts and inquired of the Lord and received from God, the greatly enhanced revelations of the creation, which he then preserved in a written text called *The Book of Moses*. This simple act of asking in faith is available to all who seek knowledge, wisdom, and answers to questions as recorded in *Matthew* 7:7; "Ask, and it shall be given you; seek, and ye shall find; knock, and it shall be opened unto you."

Had Moses been the author of the *Genesis* creation accounts, he would have written it as revealed to him in *The Book of Moses* and not as the very short accounts that we have today in the collection of manuscripts commonly called *The Bible*—the word *bible* comes from the Latin *biblio*, which means library. Moses knew that these ancient records had been passed down from Adam to him and the original record, having been written by God, were sacred and worthy of preservation, regardless of length. He merely saw to it that all these records were preserved. Each record would have been kept as a separate scroll or compiled together in one scroll of several books. When Solomon built the Hebrew temple, these scrolls were housed in a safe place within the temple confines. The collection of manuscripts commonly called *The Bible,* recognized by Protestants and Jews, was put in its final form around A.D. 100, and the invention of printing came in the 1400's. Probably the first book printed was a Collection of manuscripts commonly called *The Bible*, by Johannes Gutenberg, in 1456. Of course the Catholic church leaders tried to stop the proliferation of scripture to the common folk. Prior to this time the manuscripts were hand copied and thus easy to control. Once printed on the press, however, the task became more formidable. Leaders tried buying up the copies and burning them, burning those who had copies, and even putting to death those who had memorized scripture. The act of buying the copies had an unexpected effect. It financed the printing of even more copies of the bible to new readers. Many consider to this day that the printing press was the greatest invention of all time, as it wrested the common man from the grips of ignorance and domination by the Holy Roman Empire.

The Rod of God was lifted by Moses toward heaven and caused the Lord to send thunder and hail, mingled with fire upon the ground to smote the Egyptians.[20] It was used to smite the dust of the land, that caused the plague of lice.[21] It was this rod that turned the waters into blood. [22] The plague of frogs was caused by the rod when stretched forth over the streams, rivers, and ponds.[23] The plague of locusts was caused when Moses stretched forth his rod over the land of Egypt.[24] The Rod of God, when lifted up by Moses, parted the Red Sea for the children of Israel.[25] Moses smote the rock in Horeb, with his rod, to produce water for the children of Israel to drink.[26] In battle with the Amalekites, Moses held up the rod in his hands and eventually Israel prevailed and defeated them, under the command of Joshua.[27] Another reference to this rod, in *The Book of Jasher*, states that Moses smote Og, king of Bashan, with a stick at the ankles of his feet and slew him.[28] Og, was one of the offspring of the giants. Aaron's rod brought forth buds, bloomed blossoms, and yielded almonds as a sign from God that the tribe of Levi would minister as priests before the Tabernacle of Witness for the children of Israel.[29] The last account of this sapphire stick is that it was one of the three items placed in the Ark of the Covenant.[30] In *Maccabees* 2:1-8, it states that the prophet Jeremiah, being warned of God, commanded the Tabernacle and the Ark to go with him, as he went forth into the mountain, where Moses climbed up, and saw the heritage of God. And when Jeremy came thither, he found a hollow cave, wherein he laid the tabernacle, and the ark, and the altar of incense, and so stopped the door. And some of those that followed him came to mark the way, but they could not find it. Which when Jeremy perceived, he blamed them, saying, "As for that place, it shall be unknown until the time that God gathers his people again together, and receive them unto mercy." Is Mount Nebo—now in Jordan, across from the Dead Sea—the final resting place of the Ark, concealed in a cave? Did Jeremiah return for these treasures? According to local tradition, Jeremiah is buried on Devenish Island in Loch Erne in western Ireland and that he brought the Ark of the Covenant to Ireland, himself. But, that is another story.

Where are the books of Enoch? The writings of Enoch are apocalyptic works and there is a quotation from his writings in *Jude* 14. It appears from the writings of Paul that he had information about Enoch, taken from *The Book of Enoch*, which is not contained in the *Old Testament* as recorded in *Hebrews* 11:5. This is one of the lost scriptures or is it?

There are several "*Books of Enoch*" floating around which are forgeries or counterfeit copies. However, there are two *Books of Enoch*, which could be the lost scriptures. The first book is titled *The Book of Enoch the Prophet*. This book was translated from the original Ethiopian Coptic script in 1821, by Richard Laurence, and was published in a number of successive editions, culminating in the 1883 version. The manuscript or text was found in an Ethiopian Monastery. The back cover of the book states the following: "Like the *Dead Sea Scrolls* and the *Nag Hammadi Library*, the *Book of Enoch* is a rare and important resource that was suppressed by the early church and thought destroyed."

The title of the second book is *The Book of the Secrets of Enoch*, translated from the Slavonic by W.R. Morfill and edited, with Introduction and Notes by R.H. Charles. From the back cover of this book, it states the following: " For 1200 years this book was known only to a few people in Russia. When it was finally revealed to the world in 1892, it was announced that it was a Slavonic version of *The Book of Enoch*. This was wrong. Once translated, it was found that we have an entirely different book on and about Enoch, described, by the editor, as having no less value than the other book. (Ethiopian text)."

"This work was written in Egypt and we have determined that its author or original editor was a Hellenistic Jew. The Greek original has been lost to history, but the Slavonic text somehow survived. In its original Greek form it had a direct influence on the writers of the *New Testament*. This book was also referred to by Origen and used by the Church father, Irenaneus. It was read by, and considered valuable, by the heretics of the day in addition to mainstream Christians. This may be one reason why it was excluded from *The Holy Bible*."

Portions of both *Books of Enoch* refer to the "Giants." *The Book of Giants* was another literary work, attributed to have been written by Enoch, widely read in the Roman Empire. This book may have been taken from the original books, written by Enoch. Among the *Dead Sea Scroll* texts are at least six, or more copies of the *Book of Giants*. A fragment of that work is found in the Qumran texts in the *Enochic Book of Giants*.[31]

Another fragment of the *Dead Sea Scrolls*, called the *Pseudo-Jubilees*[32] alludes to Enoch's journeying into various heavens, and knowledge of the heavenly bodies and their courses, as well as of the calendar.[33]

The scribes of the *Dead Sea Scrolls* may have had access to one of the many copies of the original *Book of Enoch* and preserved it in their manuscripts. The same could be said about both the Ethiopian and Slavonic versions of the book. Their "writers" were probably just editors working from scribed manuscripts passed down. It is interesting to note that one editor was from Egypt, where copies of Enoch's writings were once housed in libraries of the kings and Pharaohs. Because the Qumran fragments are almost identical in context to these two *Books of Enoch*, it certainly adds credence that these books are some of the lost scriptures.

Enoch wrote 366 books or manuscripts with views on creation, eternity, the multiple heavens and their hosts, astronomy with relation to times, seasons and the calendar, and prophecy. All the above books and writings contain just that. The authors have referenced these books in many of the chapters, in our under-taking, of the writing of this book.

Twenty years after Jerusalem fell in 1099 to Christian knights, called the First Crusade, nine French knights obtained permission from King Baldwin 11 of Jerusalem to establish a home on Temple Mount under the pretext to protect Christian pilgrims traveling the roads from Europe to the Holy Land of Palestine. They were given a plot of land, by the King, which included the basement of the Temple, which was used for a stable. Thus, was the origin of their name, the Poor Knights of the Temple. For seven years these knights worked tirelessly excavating the Temple basement and foundation ruins, scarcely ever venturing outside to protect the roads. In the 1970's, Israeli archaeologists found one of their tunnels.[34] They apparently found what they were looking for, reputedly, "flying scrolls."

Remember these "flying scrolls" flew out of Adam's hands, when he was expelled from the garden, and contained all of God's knowledge of his works of creation, manuscripts called *The Divine Book of Wisdom, The Book of Adam*, or *The Book of the Angel Rezial*.

The "old ritua" of freemasonry relates that the two great pillars placed in front of the Temple of Solomon had been hollow. Inside them had been stored the "ancient record" and the "valuable writings" pertaining to the ancient past of the Jewish people. In fact, Emulating Working, an English system of Freemasonry, states the following of the pillars: "They were built to be hollow so that they could hold the archives of Masonry, and indeed the scrolls of the constitution were laid in them."

The York Rite ritual, which is commonly observed in the United States, reaffirms the above statement with the following: "They were built to be hollow in order to preserve the archives of Freemasonry from earthquakes and floods." The pillars are also referred to as the pillars of Enoch and the pillars of Seth.

The knights returned back to France, in 1126, and in 1128 persuaded the Church to back the founding of the Order of Knights Templar. In a few years the Templars became the richest and most powerful order in Europe and the source of much of their wealth was a well- kept secret. Many have surmised that they were looking for the *Dead Sea Scrolls* or perhaps the Ark of the Covenant, which incidentally disappeared long before the knights arrived in Jerusalem. The Templars maintained a shroud of secrecy of what they found in Jerusalem and thereafter they have been associated with searching for a mystery object or some ancient knowledge, which persists today.

The mystery object was indeed ancient knowledge, particularly the alchemy part of the book, which divulged the secret of turning ordinary base metals into gold. The Templars regarded the Pope as their master, which was in conflict with the King of France, Philip the Fair (1265-1314). The Order was unpopular with the King who regarded them as heretics, but at the same time he coveted their wealth for the expansion of his kingdom and devised a plan to usurp the Templars' wealth for himself. He made formal charges of heresy, spitting on the cross, homosexuality, and worshipping a devil called Baphomet, in a dispatched communiqué to the Catholic Pope, and demanded the arrest of all the Templars.

It seems that the knights near Rennes-le-Chateau, were forewarned of impending arrest and managed to avoid the trap and escaped with vast amounts of treasure. The day before the Templars arrest on October 12, 1307, eighteen ships sailed out of La Rochelle, the Templars' port and vanished from history. Philip the Fair failed to seize the wealth of two major strongholds near to Rennes-le-Chateau, Bezu and Blanchefort.[35]

At least 5000 knights were imprisoned, and their land and remaining wealth was seized by the French King. Many of the knights died under terrible medieval torture, including Jacques De Molay, the Grandmaster of the Order. Some Templars managed to escape and find safe havens in Scotland and England where they laid the foundation of Freemasonry. The first Freemasonic Grand Lodge of the World was founded in 1717, in England, and was the model for all subsequent lodges in both England and America.[36]

For the first 100 years of United States history, the government, in all three of its branches, was monopolized by Freemasons. The new government resembled one huge Freemasonic Lodge. Some high ranking Freemason officials in government were, George Washington, John Adams, Thomas Jefferson—although his activities seem to indicate he was a Freemason there is no hard evidence that Jefferson was an actual Mason—James Madison, and James Monroe. Even to the middle of the twentieth century, the Congress remained solidly Freemasonic and up to 1924 the Senate membership was 60 percent Freemason.[37]

The rites and ceremonies practiced by the Freemasons incorporate knowledge taken from the ancient *Book of Divine Wisdom.* This knowledge influenced Freemason, Thomas Jefferson, as the principal author of the Declaration of Independence. Of the 56 signers of the document, 50 were Freemasons, including the Grandmaster John Hancock. The Seal of the United States of America was originally designed by Freemason William Barton and eventually adopted with modifications. Over the pyramid part of the seal are thirteen letters, Annuit Ceoptis, meaning: "He (God) hath prospered our beginning" and written upon its base is the inscription Novus Ordo Seclorum, "New World Order." Our

forefathers of this great nation, all Freemasons, were responsible for placing the "all seeing eye" (all seeing eye of God) pyramid symbol on the dollar bill currency.[38] The important thing to remember is that our nation was founded on the divine principal of democracy, a teaching found in *The Divine Book of Wisdom*, a teaching ordained by God that inspired our forefathers. The knowledge from this ancient record also inspired European Freemason kings, nobles, and intellectuals to become involved in the creation of the Protestant Confederacy. King Frederick, of Prussia, became a high initiate of Freemasonry and a zealous supporter of the Confederacy. Philip Egalite, the Duke of Orleans of France, and Grandmaster of the Grand Lodge of France, with other French nobles, helped found a branch of the Confederacy and then proceeded to lay the foundation for the ensuing democratic revolution in France.

In Eastern Europe, King Rudolph 11, of Bohemia, a Protestant participant in the Confederacy, and his chief advisor, Pisteria, a Master Kabbalist, promoted the concept of a free Europe by organizing summits in his country that promoted democracy. In England, the Freemasons Sir Walter Raleigh, Sir Francis Bacon, and other Freemason statesmen, met to initiate affirmative action for the creation of democracy and freedom. The Rosicrucian, John Dee, advisor to the English Protestant Queen Elizabeth, persuaded the queen and much of the British government to join in on the ranks of the Confederacy.[39] Martin Luther was a fringe member of the Rosicrucians. The Rosicrucians and Freemasons worked tirelessly to create a "New World Order" based on democracy.

George Washington ascended to the degree of Royal Arch, one of the highest of Master Mason degrees. He was also honorably inducted into the Mystics of Wissahickon and the American Supreme Rosicrucian Council, which was instrumental in drafting both the Declaration of Independence and the Constitution. Freemasonry gave rise to the Rosicrucians, considered a splinter group or elite branch of the Freemasons that emerged around the time as rose windows appeared in the gothic cathedrals throughout Europe. When Washington died, the Washington monument was erected in his memory, a 600 foot Egyptian obelisk, its cornerstone laid, in a special ceremony, using the symbolic tools of Freemasonry, the square, level and plumb.[40]

The Freemasons and Rosicrucians, Thomas Jefferson and Grandmaster Ben Franklin, were adept in knowledge of the sciences, mathematics, astronomy, botany, geography, politics and law. Franklin had membership in the Appollonian Society, an esoteric fraternity founded upon the principles of ancient Egyptian wisdom. [41] Wisdom, no doubt, which source originated from the same sacred book.

Joseph Smith, the Mormon prophet, ascended to the thirty third degree of Freemasonry. He incorporated ancient knowledge, taken from *The Divine Book of Wisdom*, and made it a part of the sacred rites and ceremonies used in the temples of the Church of Jesus Christ of Latter Day Saints. The square and the compass accompany the flying angel blowing the trump of God atop the original Nauvoo Temple.

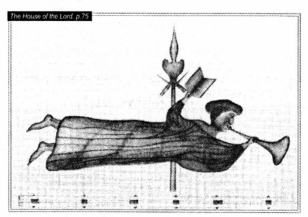

The House of the Lord, p.75

Books have been written on the possible sources of the ancient Egyptian wisdom. Some have theorized the Atlantian culture or aliens from another planet brought this wisdom to Egypt, although in a sense, Noah, his family, and all others aboard the ark, including the animals, were aliens from another dimension. The ancient writings of Enoch and *The Zohar* say otherwise. According to their writings, *The Book of Divine Wisdom*, consisting of many manuscripts, was passed down from God to Adam, from Adam to Noah, and from Noah to his children, who passed the scribed copies of the manuscripts to their children. Ham's descendants were the Egyptians. Parts of the sacred book were available to the educated class of Egypt. From these manuscripts the Egyptian mathematicians learned the concept of advanced mathematics. They learned how to employ the concept of pi, how to square numbers, how to distribute load bearing weight in architecture, learned the principles of geometry, the concept of phi, and used advanced numbering systems. There is evidence that they learned an alchemy process of electroplating gold. They learned medical procedures, such as trepanning, a surgical procedure of boring holes into the scull to relieve pressure on the brain. These ancient manuscripts contained knowledge of astronomy, the earth's geology, how to measure time, how to create, surveying, and the art of magic. The list goes on as to the knowledge contained in these manuscripts, after all, it was the handbook of God's knowledge. This knowledge, once mastered, enabled the Egyptians to build the pyramids, erect temples with amazing feats of skill, and plan city complexes with running water, sewage systems, and hot baths. The authors, however, do feel that *The Divine Book of Wisdom* and Enoch's writings may not be the entire source of wisdom possessed by the ancients. We will address this issue in the chapter titled, *The Lost Civilization of Atlantis*.

Among the Egyptians were societies that practiced magic and divination to the point that they were masters of the craft. All these skills were learned, by the magicians, from the ancient texts. According to the *Westcar Papyrus,* during the time of the Pharaoh Khufu, one magician, known simply as "Djedhi," gained legendary renown for successfully reattaching severed heads to animals.[42]

Among the Hebrews, this knowledge or secret wisdom, had been passed down from Abraham to King Solomon and taught in their School of the Prophets. Moses, no doubt, was a contributor of knowledge to this school, having learned much in the Egyptian universities. Supposedly at one time some very valuable records were transferred from the Sun Temple at Heliopolis to the School of Prophets, perhaps, under Moses' order. Later these records were carried to the eastern parts of Tibet, maybe to a monastery, and supposedly continue to exist there in a hidden underground vault.[43]

The knowledge of flight may have been learned from the ancient texts by the Hebrews. According to an ancient Ethiopian text, the *Kebra Nagast, or Glory of The Kings,* Solomon possessed some kind of airship which he would fly great distances. The writing states that Solomon had a "heavenly car" which he inherited from his forefather (King David) and used frequently. "The King.... and all who obeyed his word, flew on the wagon without pain and suffering, and without sweat or exhaustion, and traveled in one day a distance which took three months to traverse (on foot)," says the *Kebra Nagas,*[44] translated by Sir E.A. Wallis Budge, 1932, Dover, London. According to the Russian-American explorer, Nicholas Roerich, there are Central Asian legends of King Solomon journeying in an airship to Tibet.[45]

Obviously, this knowledge was lost when Solomon's temple was destroyed. According to *Genesis* 13:2, Abram was very rich in cattle, in silver, and in gold. Where did he get the silver and gold? Did he use the knowledge contained in the sacred manuscripts to produce these precious metals from ordinary base medals? The Kabbalah texts suggest that he did. In the final stanza of *Sefer Yetzirah*: "When Abraham…looked and probed…he was successful in creation. ." This passage clearly suggests that Abraham created things using methods found in the *Sefer Yetzirah*. More scriptural evidence is provided in *Genesis* 12:5, which states: "Abraham went as God had told him, and Abraham took the souls that they had made in Haran." According to some Kabbalah teachings, this indicates that Abraham actually used the powers of *Sefer Yetzirah* to create people or as some have interpreted created Golems, soulless beings, to do the work.[46]

Do the manuscripts of *The Divine Book of Wisdom* still exist? There is much evidence that suggests they indeed do exist today, or at least scribed copies of the original scrolls. Remember the eighteen ships that escaped from La Rochelle with the Templar's treasure? It appears that at least one of the ships sailed to Scotland where some of the Templars took refuge in an abbey called Kilwinning, an early strong post and major centre for Templar power. The Pope had excommunicated the Scottish King, Robert the Bruce, making it a safe haven for the Knights Templars.[47]

317

Shortly afterward part of the Templar's treasure was used in the construction of the Roslyn Chapel, near Edinburgh. William St. Clair, a Knight's Templar and last Prince of Orkney, began building the tiny chapel about 1446, and it was completed sometime after his death in 1484. The extraordinary architecture on the outside bristles with flying buttresses and medieval gargoyles in the highest style of florid gothic. The chapel interior is profusely carved with biblical scenes depicting the Fall of Man, the Expulsion from the Garden of Eden, the Birth of Christ, the Crucifixion and the Resurrection. Hundreds of exquisitely carved animals, fruit, leaves, and figures cover the pillars and arches. One of the things the Knights Templars found in the ruins of Solomon's temple was a drawing of the Heavenly Jerusalem City.[48] The drawing was in the gothic style and resulted in an explosion of gothic style cathedrals being built all over Europe, by the Freemasons. Roslyn Chapel was one of these gothic structures.

William St. Clair was a Templar and Freemason. There is a carving on the wall outside Roslyn that depicts a Freemason ceremony, with the candidate blindfolded and with a noose around his neck. A Templar, with the cross on his tunic, is holding the rope to the noose of the candidate. It is believed that the building of Roslyn marks the first advent of Freemasonry in the United Kingdom.[49]

Redrawing of First Advent of Freemasonry

During the construction of the chapel, in 1447, a fire occurred that caused William St. Clair to become frantic until he learned from his chaplain that the four great trunks full of charters—could have been manuscripts or scrolls—were safe and sound, and according to the record, "he became cheerful."[50] We can say with some certainty that these trunks contained the scribed books, manuscripts, charters, or scrolls that comprised *The Book of Divine Wisdom,* perhaps other books that Enoch wrote, and maps of the world—discussed in up coming chapters. Roslyn Chapel was built as a shrine to house and protect these manuscripts, the source of the ancient knowledge of the Babylonians, Sumerians, Egyptians, Sufis, Freemasons, and Rosicrucians. Somewhere in a sealed vault, perhaps, under the floor, in a hollow column—the Apprentice Pillar—or concealed in the altar itself, hidden yet, accessible and safe, these forgotten records may still lie in that chapel. Even the name, Roslyn, in Scottish Gaelic, denotes meaning. 'Ros' means knowledge and 'linn' means generation. Translated in modern Gaelic, it would be 'ancient knowledge passed down the generations.'[51]

Copies of some, if not all, of these records probably lie in the hands of the Grand Master Masons of Freemasonry and Rosicrucians, not available to the lower order initiates. It takes years of disciplined study, devotion and probation to understand some of these texts. Most have to be translated into other languages for understanding. So why the big secrecy of these organizations? Why not share the knowledge? The answer is simple; knowledge is power and power in the wrong hands, especially this type of knowledge and power, could bring the world back to the dark ages. Democracy would no longer exist, people would be enslaved, human rights would be nonexistent, and religious freedom would cease.

The Egyptian Pharaoh Tutankhamun was a 'freemason' and holder of divine knowledge. The clues are found in his tomb as evidence such as the Freemason 'apron', his dues to the lodge of two jars of honey, and in the mummy bandages was the dagger of the Freemason Tyler. There is a life-size statue of him wearing the pyramid skirt of fertility that guards his coffin containing his sarcophagus. These items confirm his identity as an esotericist (Freemason)[52]

Previously the Freemasons could trace their 'roots' to Solomon's temple, at least to the Knights Templar but now as of this writing their 'roots' appear to go back, much further, even to Adam who gained knowledge from the 'flying scrolls', scrolls containing ancient knowledge and wisdom written by God.

There are other scribed copies of these manuscripts that perhaps still exist. The famous clairvoyant Edgar Cayce (1877-1945), also known as the sleeping prophet, identified another location where they are stored for safe keeping in the event of another world catastrophe. If such an event occurred the records could be retrieved and knowledge preserved. This site is the Sphinx of Egypt. According to Cayce, the Sphinx guards the Hall of Records, in an underground vault, located in a chamber under the front paws of the Sphinx. The descendants of Ham, most likely, put copies of these ancient records in this repository for safekeeping. These records joined other ancient records put there previously for safekeeping and will also be addressed in the chapter *The Lost Civilization of Atlantis.*

In 1990, a geologist Thomas Dobecki, sent vibrations down into the rock under the Sphinx's front paws and found sonar evidence of a rectangular underground chamber lying about 25 feet underground.[53] Another researcher, Dr. Joseph Schor of the Schor Foundation, in connection with Florida State University, also explored the Giza area with ground-penetrating radar, throughout the 1990s and early 2000s. He also detected a chamber some 30 feet under the Sphinx. It appears to be a chamber about 26 feet wide, 40 feet long, and approximately 30 feet from floor to ceiling in height.[54] In October 1992, a French engineer named Jean Kerisel was using ground-penetrating radar to explore the descending passage to the underground chamber of the Great Pyramid. His equipment detected a 'structure' that could be a corridor leading directly to the Sphinx.[55] If indeed Cayce is correct then the Hall of Records would have to have some means of access and what better way than an underground passageway from the Great Pyramid to the Sphinx. Cayce, in one of his readings, disclosed that there is a passage from the right forepaw of the Sphinx leading to the record chamber.[56] In 1976, a Japanese research team, using seismic and magnetic remote sensing, detected a possible tunnel 2.5 to 3 meters below the ground surface, near the right paw. During a full-scale clearing of the Sphinx in 1926, a passage under the rear part was resealed with masonry and cement. Two British archaeologists, in 1980, mapping the structure, discovered the sealed passage. Ancient legends as well as the *Egyptian Book of the Dead*, speak of underground chambers near the Sphinx.[57] In fact the whole Giza plateau is known to be honeycombed with underground tunnels. The problem is the Egyptian government. They have the exclusive right to explore the Sphinx for an opening to the underground chamber, but unfortunately have no plans to do so.

Edgar Cayce gives two other locations where ancient records lie hidden. One is under the waters near the Bahamian island of Bimini and the other is beneath a temple in Mayan lands in the Yucatan. Cayce states, "The records in this temple are "overshadowed" by another temple that is built over it." The ancient Mayan city of Piedras Negras located deep within the jungle lowlands of Guatemala beside the beautiful but treacherous Usumacinta River seems to be the site to which the Cayce readings refer.

According to Cayce, the three repositories of ancient record sites contain exactly "thirty-two tablets or plates", linens, gold, and artifacts relating to the cultures that created them. He indicated that mummified bodies are buried with the records. The 32 stone tablets contain the history and origins of mankind. When these records are finally located it will take time to translate language and research interpretation. Let's hope that whoever finds these records doesn't take as long as it has taken to interpret the *Dead Sea Scrolls* (45 years) found in the Qumran caves.[58]

There is another site in the Yucatan that is the possible location of these records. Votan, a grandson of Noah, built a temple by the Huehuetan river, called the "House of Darkness", wherein he deposited the national records. This information came from Nunez de la Vega, Bishop of Chiapas, who recalled burning a book purported to be written by Votan describing the erection of a Tower of Babel and the above event. Nunez de la Vega further stated Votan wrote "that he saw the great wall, namely the Tower of Babel, which was built from earth to heaven at the bidding of his grandfather Noah."[59]

The authors believe that the Yucatan and perhaps the Bimini Hall of Record sites do contain texts of *The Divine Book of Wisdom*, perhaps written on tablets of stone. This was the source of the ancient Mayans knowledge and wisdom, which was given to them by Votan, grandson of Noah. For those who would like more information on the

Yucatan and Bimini sites, the book *THE LOST HALL OF RECORDS* by John Van Auken and Lora Little, Ed. D, is excellent reading. (see also page 280 for location of Yucatan hall of records)

We discussed earlier, in this chapter, *The Book of the Angel Rezial,* which is suppose to be the actual text of *The Divine Book of Wisdom.* We concluded that perhaps the older portion of the text, in the back section of the book, could possibly be scribed text from the passed down copies of the original manuscript, however the more modern part of the text, in the front of the book, were added and altered from generation to generation.

The last writings and texts that contain some of the ancient wisdom of *The Divine Book of Wisdom* is *The Nag Hammadi Library.* These are the Gnostic scriptures, writings that appear to have been influenced by Jewish traditions and texts containing such traditions, Jewish texts that were influenced by *The Divine Book of Wisdom.* Gnosticism begins where the history of the Essenes breaks off. The Gnostics were out of the "main stream" of Christianity and were described as "heretical", but as we believe only a people ahead of their times. Of course, only portions of their writings contain truths that we have referenced to in previous chapters.

In summary of this chapter, we have traced what happened to three items that were aboard the ark, namely the Rod of God, the manuscripts or *Books of Enoch*, and the writings of *The Divine Book of Wisdom.* We have identified books that contain some of this wisdom available today. The surprise was that the first five chapters of *Genesis* are some of the manuscripts from *The Divine*

Book of Wisdom; the creation accounts. We have revealed how this divine book of knowledge inspired our nation's founding fathers and leaders of European nations to build nations on the concept of democracy, freedom, and human rights under the one true God. We have disclosed the probable locations of the whereabouts of these scribed documents and those organizations that also have scribed copies, organizations that take a vow of secrecy as from where their source of ancient wisdom derives. We have disclosed how this knowledge has affected religion from the texts of the collection of manuscripts commonly called *The Bible*, how it has been incorporated into the temples of the Church of Jesus Christ of Latter Day Saints, even influencing the Vedic texts and perhaps Asian beliefs. This knowledge has been all around us for centuries and yet the spiritually asleep could not see the forest for all the trees. For those of us spiritually awake, GOOD NEWS, there is much of these writings still awaiting discovery.

In conclusion of this chapter, just remember that just maybe somewhere in the basement of a dusty dark museum, or hidden in stone boxes in some remote cave or Yucatan temple, maybe in a Tibetan monastery resting high in the Himalayas, perhaps in a yet undiscovered archeological find, could be in your great grandparents trunk in your attic, more likely in some-one's private collection of ancient artifacts, or in a sealed chamber of the Sphinx or Roslyn Chapel, more of these ancient manuscripts will one day be found. Manuscripts that will rewrite the history of the world and the origin of man and restore ancient knowledge.

End Notes

[1] *(The Book of Jasher, et al pp.17, 18,19)*

[2] *(The Book of Jasher*, et al pp.65,75)

[3] (*Mormon Doctrine*, Bruce R. McConkie, 1966, Publishers Press, Salt Lake City, pp.563-564)

[4] (Avodah Zarah 14b. CF. Barceloni. p. 100)

[5] *(The Book of Abraham* 1:23-25)

[6] *(The Book of Jasher* et al p. 221)

[7] *(The Book of Jasher* et al p. 221)(Exodus 4:20)

[8] *(The Book of Jasher* et al p. 200)

[9] (Taken from **The New Complete Works of Josephus © 1999** by **Kregel Publications**. Published by Kregel Publications, Grand Rapids, MI. Used by permission of the publisher. All rights reserved Book 2, Chapter 9:6)

[10] (Taken from **The New Complete Works of Josephus © 1999** by **Kregel Publications**. Published by Kregel Publications, Grand Rapids, MI. Used by permission of the publisher. All rights reserved Book 2 Chap. 9:6)(Acts 7:20)

[11] (*The Book of Jasher*, et al pp.212,224)

[12] *(The Complete Works of Josephus*, et al Book 2 Chap. 10:1, 2)

[13] *(The Complete Works of Josephus*, et al Book 2 Chap. 10:2)

[14] (The *Book of Jasher* et al pp.209,210)

[15] (KJV Acts 7:22)

[16] (*The Book of Jasher* et al pp.223,224)(Exodus 7: 9-12)

[17] *(The complete works of Josephus*, et al, p. 103)

[18] *(Pearl of Great Price, Moses* 6:8-9)

[19] *(Mormon Doctrine* pp.563- 564)

[20] *(Exodus* 9:23)

[21] *(Exodus* 9:16)

[22] *(Exodus* 7:17)

[23] *(Exodus* 8:5)

[24] *(Exodus* 10:13)

[25] *(Exodus* 14:16)

[26] *(Exodus* 17: 5-6)

[27] *(Exodus* 17:9-13)

[28] (The *Book of Jasher*, et al p. 240)

[29] *(Numbers* 17-18)

[30] *(Hebrews* 9:4)

[31] (4Q532)(*The Dead Sea Scrolls Uncovered*, Robert Eisenman & Michael Wise, 1992, Penquin Books, N.Y., p95)

[32] et al (4Q227)

[33] (*The Dead Sea Scrolls Uncovered*, p. 96)

[34] *The Atlantis Blueprint*, ©2000 by Rand Flem-Ath and Colin Wilson. Published by Delacorte Press a division of Random House, Inc., NY, NY. p. 220)

[35] (*The Atlantis Blueprint*, et al p. 226)

[36] Excerpts taken from *The Return of The Serpents of Wisdom* ©1997 by Mark Amaru Pinkham. Published by Adventures Unlimited Press, Kempton, ILL. All rights reserved. Used by permission of the publisher, pp.270,280)

[37] (*The Return of the Serpents of Wisdom*, et al p. 291)

[38] (*The Return of the Serpents of Wisdom*, et al pp.292, 293)

[39] (*The Return of the Serpents of Wisdom*, et al pp.286, 287)

[40] (*The Return of the Serpents of Wisdom*, et al p. 294-295)

[41] (*The Return of the Serpents of Wisdom*, et al p. 295)

[42] (*The Return of the Serpents of Wisdom*, et al p. 205)

[43] (*The Return of the Serpents of Wisdom*, et al p228)

[44] (*The Queen of Sheba and Her Only Son Menyelek* (*Kebra Nagast*)

[45] *Lost Cities of Atlantis, Ancient Europe & The Mediterranean,* ©1996 by David Hatcher Childress. Published by Adventures Unlimited Press, Kempton, ILL., p. 440)

[46] (Excerpted from *Sefer Yetzirah* by Aryeh Kaplan ©1997 with permission of Red Wheel/Weiser, York Beach, ME and Boston, MA, Introduction, p. xiii)

[47] (*The Atlantis Blueprint*, et al p. 282)

[48] (*The Atlantis Blueprint*, et al p. 265)

[49] (*The Atlantis Blueprint*, et al p. 271)

[50] (*The Atlantis Blueprint*, et al p. 282)

[51] (*The Atlantis Blueprint*, et al p. 272)

[52] (*The Tutankhamun Prophecies*, Maurice Cottrell, 2002, Bear & Company, Rochester, Vermont, pp.225, 226)

[53] (*The Atlantis Blueprint*, et al p. 107)

[54] (*The Lost Hall of Records*, John Van Auken and Lora Little, Ed.D., 2000, Eagle Wing Books, Inc. , p. 33)

[55] (*The Atlantis Blueprint*, et al p. 107)

[56] (reading 378-16, October 29, 1933)

[57] (*Mysteries of Atlantis Revisited*, Edgar Evans Cayce/ Gail Cayce Schwartzer/ Douglas G. Richards,1988, St. Martin's Paperbacks, N.Y., N.Y., pp.142,153-155)

[58] (*The Lost Hall of Records*, pp.19, 22, 35)

[59] *Atlantis Mother of Empires,* ©1939 by Robert B. Stacy-Judd. Published by Adventures Unlimited Press, Kempton, ILL. p. 98)

E. J. Clark & B. Alexander Agnew, PhD

The Giants

The Greek poet Homer wrote in 400 B.C., "On the earth there once were giants."

Genesis 6:4 states: "There were giants in the earth in those days (before the flood); and also after that (after the flood), when the sons of God (fallen angels) came in unto the daughters of men, and they bare children to them."

Who were the giants spoken of by the above writings? They were the offspring of the Nephilim and the daughters of men. The Nephilim were the fallen angels, cast out of heaven after rebelling against God. The word Nephal means, "to fall" and means "the fallen ones." Their leader was also a fallen Son of God, Lucifer (Satan). The Nephilim came from a class of angels called "the watchers", and not all watchers fell. The souls derived from the union of the Nephilim and mortal women were called the "mixed multitude" and consisted of five categories: Nephilim, Gibborim, Anakim, Rephaim and Amalekites. All giant tribes were of one of these categories or of one of a branch of these categories which were of different names. (Readers are referred back to the chapter, *The Garden of Eden*, for more on the divisions of fallen angels)

When these angels were cast out of heaven, they came to both the spiritual and temporal creations. It was Lucifer who deliberately tempted Adam and Eve, in the Garden of Eden, in an act of rebellion against God for expelling him from heaven. Satan, along with the fallen angels, wanted to gain control of both creations and create worlds or a home for themselves. In order to gain control, Adam had to lose dominion. It was a plan that backfired on Lucifer. After Satan beguiled Adam and Eve and they were cast out of the garden, God said to the couple that the seed of the woman would one-day defeat Satan and restore man's dominion.

Genesis 3:15 records this promise in the following scripture:

"And I will put enmity between thee and the woman, and between thy seed and her seed; it shall bruise thy head, and thou shalt bruise his heel."

In order to avoid this predicted defeat it would be necessary to find some way to prevent the seed of the woman from being born into the world. So, a plan was conceived by Lucifer and his followers to corrupt the pure Adamite line and make it impossible for the promised Redeemer to be born through the seed of the woman. (Eve) The plan would be accomplished by sending the Nephilim to marry the daughters of men and produce the giant half-human half-demonic nations through them, which in turn would corrupt all God's creation, thereby preventing the coming of the Redeemer. The plan included teaching the giant offspring all manner of evil to inflict on men and to implement the worship of idols, which was the worst insult of all to God.

During the first irruption of the Nephilim on the spiritual creation, the plan almost succeeded, for all flesh had been corrupted upon the spiritual earth except for Noah and all those preserved on board the "Ark of Millions of Years." *Genesis* 6:4 confirms the purity of Noah and his lineage in the following scripture: "Noah was perfect in his generations."

The flood was sent as a judgment against Lucifer and the Nephilim. Since the whole spiritual creation was corrupt except for Noah and his entourage, God destroyed the giants and all living things on the "airy" globe by a universal flood, save for those on board the ark. He then removed or "flung" the flooded spiritual creation through millions of years of time, dimensions and space to its carbon copy and soul mate, the temporal earth. Noah and his entourage were transported to this planet so that the pure Adamite stock could be preserved so as to guarantee the coming of the seed of the woman.

Lucifer's plan however, extended to the temporal creation as well. The Nephilim may have been on this earth as early as 50,000 years B.C., maybe even earlier. They married the daughters of men on this planet and produced giant off spring that were corrupting all flesh in the same manner that occurred on the spiritual creation. In order for Satan's plan to work they had to close the "loop holes" and corrupt all men from both creations.

Ancient records give us a description of the unmatched intelligence of these beings. The knowledge that they consciously brought to the earth was generations ahead of the present occupants. They taught the humans living on the earth about aesthetics, music, engineering, and war. They also took the pure and simple process of procreation and corrupted it. What was once a sacred event between two married adults became the center of attention for the planet. By the time they were done with their mission to destroy the family relationship, men could barely remember what the true relationship between husbands and wives should be. The lack of family integrity was so widespread it could be argued that God's mission for the earth and the creation of man had been completely replaced with Satan's plan. No child, no woman, no young man was unaffected by the deliberate destruction orchestrated through the giants in the land.

The 9,500 B.C. cataclysm of this planet was sent by God to destroy the giants and all their corruption of living things here. All the giants were destroyed but there were human survivors of the cataclysm. These survivors were pure and genetically uncorrupted.

The flood, destruction of the spiritual creation, and cataclysm however failed to stop Lucifer from making a further attempt to prevent the coming of the promised Redeemer. Since God had promised never again to send a flood to destroy the world, Lucifer reasoned that it was safe enough to attempt a second try to once again corrupt the surviving seed of men by marriage to their daughters and produce another race of giant offspring. A second group of Nephilim was sent to accomplish the task. They came as early as conditions on this planet permitted following the cataclysm. The Lost Civilizations chapters of this book will go more into detail of their occupation of this planet.

Once again, the unions produced giants and races of them eventually occupied the land of promise, where the seed should be born. They were here well in advance of Noah and had established mighty cities by 8,000 B.C. Jericho is the world's oldest town, dating back more than 10,000 years, and it was built by the Hittites. There were large numbers of giants living in the world; not just a few but many tribes. Records, skeletons, and other archeological evidences have been discovered in America, Mexico, South America, Europe, China, Africa, and Australia.

In the area of the Promised Land alone, there were twenty tribes of giants spoken of in *The Bible*. Besides being giant in size, they had several different identifying genetic traits that set them apart. Most had 6 fingers on each hand and 6 toes on each foot. Some would only have the 6 toes on each foot and normal hands and others just the reverse. A few tribes even had double rows of teeth. Skull volumes indicate they had a brain chamber of at least 3200 cubic centimeters.

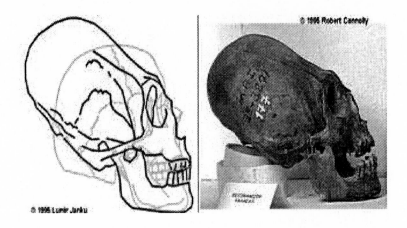

Skull of giant compared to human

This is clear evidence that their brains were about 3 to 6 times the average human brain size. Hundreds of examples of these beings have been exhumed and bear witness of their prominence in human society. They were truly giant in that their body parts were in proportion to their height. Besides possessing height, their physiques were robust and they had enormous strength. Evidence shows that this second generation also possessed tremendous knowledge and drove mankind to commit murder for profit. They were so powerful and cunning that ordinary men could not withstand them. Reconstructive palaeobiologists have estimated their body weight from 350 to 950 pounds, depending on height. They were fast, extremely intelligent and had advanced technology which attracted an unwavering reverence from mortal men and women. Modern archeology supports the theory that men were widely enslaved and manipulated by giants.

In the beginning the first generation giant offspring were around 36 feet tall, later generations gradually diminished in size to about 10 to 18 feet tall as being the average size of biblical times. After biblical times their size further diminished to about 7 to 12 feet tall. Compared to the average 5 foot six inch size of men, giants and their offspring were undoubtedly at the top of the food chain.

Below is a list of the giants and scripture documentation of where they are listed in *The Bible*.

Anakims- DEU. 1:28; 2:10,11,21; 9:2; JOS. 11:21, 22; 14:12,15.

Amorites-GEN. 14:7, 15:16,21; EX. 3:8,17; 13:5; 23:23; NUM. 13:29; 21:13,21, 25,26,29,31,32,34; 22:2; 32:33; DEU. 1:4, 7,19,20,27,44; 3:2,8,9; 3:28,9; 4:46,47; 7:1; 20:17; 31:4; JOS. 2:10; 3:10; 5:1; 7:7; 9:10; JOS. 10:5,6,12; 12:2,8; 13:4,10,21; 24:8,11; 24:12,15,18; JUD. 1:34,35,36; 3:5; 6:10; 10:8,11; 11:19,21,22,23; 1. SAM. 7:14; 11. SAM. 21:2; 1. KING 4:19; 9:20; 21:26; 11. KING 21:11; 11. CHR. 8:7; EZRA. 9:1; NEH. 9:8; PS. 135:11, 136:19.

Avims-DEU. 2:23

Canaanites (which included the Philistines)-GEN. 10:18,19; 15:21; 24:3,37; 34:30; 50:11; EX. 3:8,17; 13:5,11; 23:23; NUM. 13:29; 14:25,43,45; 21:3; DEU. 1:7; 7:1; 11:30; 20:17; JOS. 3:10; 5:1; 7:9; 12:8; 13:4; 16:10; 17:12,13,16,18; 24:11; JUD. 1:1,3,4,5,9,10, 17,27,28,29,30,32,33; 3:3,5; 11. SAM. 24:7; 1 KING. 9:16; EZRA 9:1; NEH. 9:8,24; OBA. 20.

Caphtorims-DEU. 2:23.

Emins-GEN. 14:5; DEU. 2:10.

Gibborim-11. Sam. 23:8-39; 1King. 1:8; 1. CHR. 11:9-47, 29:24.

Hittites-GEN. 15:20; EX. 3:8, 17; 13:5; 23:23; NUM. 13:29; DEU. 7:1, 20:17; JOS. 1:4, 3:10, 12:8, 24:11; JUD. 1:26, 3:5; 1. KING. 9:20, 10:29, 11:1; 11. KING. 7:6; 11. CHR. 1:17, 8:7; EZRA. 9:1; NEH. 9:8.

Hivites-EX. 3:8,17; 13:5; 23:23; DEU. 7:1; 20:17; JOS. 3:10; 9:7; 11:19; 12:8, 24:11; JUD. 3:3,5; 11. SAM. 24:7; 1KING. 9:20; 11. CHR. 8:7.

Horims-DEU. 2:12,22.

Kadmonites-GEN. 15:19.

Kenites-GEN. 15:19; NUM. 24:21; JUD. 4:11; 1. SAM. 15:6, 27:10, 30:29; 1. CHR. 2:55.

Kenizzites-GEN. 15:19.

Nephilim-GEN. 6:4; 14:5,6; 15:19-21; EX. 3:8, 17,23; DEU. 2:10-12, 20-23; 3:11-13; 7:1; 20:17; JOS. 12:4-8; 13:3; 15:8; 17:15; 18:16.

Perizzites-GEN. 15:20, 34:30; EX. 3:8, 17; 23:23; DEU. 7:1; 20:17; JOS, 3:10, 12:8, 17:15, 24:11; JUD. 1:4,5, 3:5; 1. KING. 9:20; 11. CHR. 8:7; EZRA. 9:1; NEH. 9:8.

Raphaims-GEN. 14:5; 15:20.

Zamzummims-DEU. 2:20.

Zebusites-Gen. 15:21; EX. 3:8,17; 13:5; 23:23; NUM. 13:29; DEU. 7:1; 20:17; JOS. 3:10; 12:8; 15:63; 24:11; JUD. 1:21; 3:5; 19:11; 11. SAM. 5:6,8; 1. KING. 9:20; 1. CHR. 11:4,6; 11. CHR. 8:7; EZRA. 9:1; NEH. 9:8.

Zuzims-GEN. 14:5.

The first generation offspring of "fallen angels" were super intelligent beings having learned much of their knowledge from their fathers, who having been in the presence of God knew much of his workings. Later generations became power-hungry ruthless rulers and despots. They, and their offspring, through tyranny, became "mighty men." Their rulers and kings proclaimed themselves to be "sons of God" in order to enhance their power and prestige. In a sense they were correct in proclaiming themselves "sons of God" but failed to acknowledge they were cast out of heaven and were the fallen ones with Lucifer as their leader.

Perhaps the most famous giants in *The Bible* were Goliath and Og, the Amorite king of Bashan. Goliath, a Philistine, was from Gath. He had four relatives that were also giants:

1. Ishbi-Benob whose bronze spearhead weighed more than seven pounds

2. Saph or Sippai who died in the battle of Gob (1 Chron. 20:4)

3. Lahmi, the brother of Goliath the Gittite

4. An unnamed huge man with six fingers on each hand and six toes on each foot.

Goliath and these four were all descendants of Rapha in Gath, "and they fell at the hands of David and his men."[1] Apparently not all giants were deemed forbidden company because King David employed the Gibborim or "mighty men" as his body and palace guards. Some were even made captain over his armies. When David was a fugitive, he gathered around him six hundred Gibborim for protection. According to the scriptures these "mighty men" were loyal and valiant servants of King David. The word **gibberish** in our language is a reflection of the speech of the Gibborim. (refer to Gibborim for scriptural references) David was one of the most powerful kings in history. He is one of the few kings who had true divine endorsement. This endorsement is perhaps the only thing on the planet that was ostensibly more powerful than the rule of giants.

In order to stop the corruption of the seed, the Hebrews forbade marriage outside their race and forbade bestiality. It was the reason that God commanded the Israelites to kill the giant races including the women and children.

There is an old rabbinic tradition which alleges that the fallen angels were attracted to the beautiful long hair of women, in so much that in *1 Corinthians 11:10*, Paul instructs that women should cover their heads " because of the angels."

The Zohar comments on the life span of the giants as follows: "They lived to a great age until at last half their body became paralyzed while the other half remained vigorous. They would then take a certain herb and throw it into their mouths and die, and because they thus killed themselves they were called Refaim." According to Rabbi Isaac's comments in *The Zohar*, he reports that they used to drown themselves in the sea, as it is written, "The Refaim are slain beneath the waters."[2]

All of the writings above are accounts of the giants who lived in the "Land of Promise". There is worldwide evidence of other tribes of giants that once existed on the earth. Let us begin with the following discoveries in America:

1833. Soldiers digging a pit for a powder magazine at Lompock Rancho, near San Luis Obispo California, unearthed a twelve foot skeleton that had a double row of upper and lower teeth. Surrounding the skeleton were huge stone axes, carved shells, and blocks of porphyry covered with unintelligible symbols. The local authorities ordered it secretly reburied because the local Indians began to attach religious significance to the artifacts.

1880. Clearwater, Minnesota, several giant skeletons were found with double rows of teeth.

1886. Ellisburg, Pennsylvania, an eight-foot skeleton was discovered.

1886. New York, dozens of human skeletons were discovered with oddly shaped skulls averaging seven feet in height.

1886. Illinois, Logan County, a number of large skeletons and artifacts were found.

1887. Wisconsin, Le Cresent, bones of giant humans were discovered in burial mounds.

1891. Crittenden Arizona, a giant skeleton was unearthed along with a huge stone coffin that had apparently once held the body of a twelve-foot tall man. A carving on the granite case indicated that he had six toes.

1923. Grand Canyon Arizona. . the bodies of two petrified human beings, fifteen and eighteen feet in height, were discovered.

1947. Death Valley California. . the skeletons of nine-foot tall humans were found.

On Catalina Island, California, some time in the late 19th century, giants with double rows of teeth, red hair, and skulls three to six times the size of normal humans were discovered.

Buffalo Bill Cody spoke of Sioux Indians who told about giants that ran down the buffalo.

The oral traditions of the Delaware and the Sioux Indians tell of a race of "great stature" but cowardly with whom they entered into conflict. The Allegheny River and Mountains are named after the Allegewi, a giant tribe, but the Iroquois Confederacy drove the giants out of their strong, walled cities, and the Sioux finished them off when they attempted to relocate in what is now Minnesota.

The Washoe Indians have a legend of having a battle with a tribe of giants before the coming of the Piutes. The Washoe Indians, upon first coming to the lake, discovered a dozen camps where many giants lived. The Indians attacked and the giants threw huge boulders at them. After a fierce battle, many giants were killed, and the others fled. With the giants gone, the Piutes then moved north to Pyramid Lake from Walker Lake where they lived in peace with the Washoe Indians.

There are also legends of a race of redheaded eight-foot cannibal warriors called the "SI-TE-CA" by local natives. They allegedly inhabited the southwest and their remains have been found at Lovelock, Nevada in 1911-1912. Two of the skulls are now in museums in Lovelock and Winnemucca.

According to Native American legends, the giants ate people, a trait exhibited in the spiritual creation. The Midwest giants had double rows of upper and lower teeth, some also had elongated canine teeth. The American giants exhibited six digits on both feet and occasionally six digits on both their hands as well.

The jaw of a giant compared to a human jaw

The authors wonder if any museums, besides the Lovelock-Winnemucca museums, contain any of these skeletal remains or just exactly what has happened to them? Some were re-interred but we imagine it would be interesting to do a DNA profile on the bones if any could be located. Why hasn't more been said about these findings? Why has history remained silent on the evidence found?

Most all legends declare that the megalithic cities, megalithic structures, cyclopean and polygonal type of walls were built by an ancient race of giants that once inhabited the entire earth.

According to a story written in the 12th century by Geoffrey of Mowmouth, the stones of The Giant's Ring were originally brought from Africa to Ireland by a race of Irish giants. The stones were located on "Mt. Killaraus" and were used as a site for performing rituals and for healing.

Centuries later, in 480 A.D., the legendary Merlin the magician was able to magically transport these stones from Ireland to the present location now known as Stonehenge. Merlin accomplished this feat by the manipulation of the "Life Force" energy that caused the giant stones to fly through the air. Legend has it that the stones were moved to create a war memorial over the site of several hundred graves believed to be slain Saxon soldiers, at the request of King Aurelius of Briton. In ancient times Stonehenge was called Chiorgaur, which means dance of giants.

Today it is known that the Saracen stones or outer ring of stones of Stonehenge were quarried from Marlborough Downs, some twenty miles away. It is believed that most of the eighty blue stones, weighing four tons each, were quarried from the rock of the Presely Mountains of South Wales. From whatever source the huge stones came, the giants were the only ones capable of building this structure. As previously mentioned in the chapter, *The Union of The Polarity*, Stonehenge was built on a powerful vortex and built in three stages. It was by utilizing the "Life Force" energy that the great stones were quarried and transported. Merlin may have brought some of these stones from Ireland to finish the structure.

The Celtic giants were descendants of the Cimbri or Cimmerian giants that had migrated into northern Europe. Of course, not all Celts were giants, but there was a race of Celts that were of colossal height of nine feet or more who lived among the ordinary size Celts. Many historians describe them as "blond, blue eyed, giants" who struck terror in the hearts of their enemies.

These wandering Celts first robbed then chased smaller men off the best lands. As they multiplied they required more territory, crowding still others out. They eventually grew into some sixty different tribes, each with a different name. In Upper Asia, Asia Minor, and Mesopotamia, these plunderers were known as the Gomarian Sacae. By the time they had migrated and settled in Europe, they decided to shed their derisive name which meant "robbers' and began calling themselves the Celtae which means "potent and valiant men." The Greeks understood them as Galatai or Galates, the Spaniards called them Keltos, the Romans heard their name as Galli, and the French called them Gauls. However, it was one and the same name.

Steve Quayle has studied the races of giants for thirty years. He has an excellent web site that I have taken some of the above information from. In his web site he has this to say about the Celtic giants:

"In keeping with their Sacae tradition, these "potent and valiant men" continued to rob. But in their new land their new name, Celtae, better fit their appearance—which alone was enough to demoralize their enemies. Indeed, even the Romans, ordinarily a very brave people, came to dread the sight of them. Concerning their frightening look, historian Ammianus Marcellinus writes: "Nearly all the Gauls are of a lofty stature, fair and of ruddy complexion; terrible from the sternness of their eyes, very quarrelsome, and of great pride and insolence. A whole troop of foreigners would not be able to withstand a single Gaul if he called his wife to his assistance who is usually very strong and with blue eyes; least of all when she swells her neck and gnashes her teeth, and poising her huge white arms, proceeds to rain punches mingled with kicks, like shots discharged by the twisted cords of a catapult. The voices of most of them are formidable and threatening, alike when they are good-natured or angry."

The following is another excerpt from his web site:

"Diodorus, an ancient Greek historian, says: "The Gauls are terrifying in aspect and their voices are deep and altogether harsh; when they meet together they converse with few words and in riddles, hinting darkly at things for the most part and using one word when they mean another; and they like to talk in superlatives, to the end that they may extol themselves and depreciate all other men. They are also "tall" in stature, with rippling muscles, and white of skin, and their hair is blond, and not naturally so, but they also make it their practice by artificial means to increase the distinguishing color which nature has given it. For they are always washing their hair in lime-water, and they pull it back from the forehead to the top of the head and back to the nape of the neck, with the result that their appearance is like that of Satyrs and Pans, since the treatment of their hair makes it so heavy and coarse that it differs in no respect from the mane of horses."

Quayle's web site continues:

"Diodorus' account of the way the Celts wore their hair agrees with those representations of the Anakim giants found in the great temple of Abu Simbel. This design depicts "the king contending with two men of large stature, light complexion, scanty beard, and having a remarkable load of hair pendant from the side of the head." Besides the similarity in strange hair styles, the Celts and the Anakim both wore torque necklaces. Proof that the Anakim wore them as distinguishing emblems appears in the name. For from anaq, the Hebrew word for necklace or neckpiece, came the name Anakim, which means "People of the Necklace." That the tow-headed Celts also adorned their necks with a twisted strip of metal, usually gold or silver, is not only shown by the ancient historians but verified by many archaeological finds. The enemy of course always took note of the Celts' attention-getting neckpieces. And so did the Roman poet Virgil, in these memorable verses:

"Golden is their hair, and golden their garb. They are resplendent in their striped cloaks, and their milk-white necks are circled with gold."

The Celts loved bright colors. The Romans described them as a towering folk whose clothing was most striking; their shirts and trousers were dyed and embroidered in varied colors and in their coats were set a pattern of checks of varied hues, similar in fashion to the Scottish tartan. On their heads these Titans wore bronze helmets with horns attached.

These giants practiced human sacrifice to appease the gods and several early historians portrayed most Gallic males as homosexuals.

In about 387 B.C., one of the southernmost Celtic tribes called the Senones, numbering some three hundred thousand, having over populated their region, crossed over the Apennines and swarmed into northern Italy, attacking and looting Etruscan towns. The Celtic Senones, liking the country side of northern Italy, decided to occupy it and oust its former Italian residents. The Italians (Etruscans) were unable to halt the onslaught of the Senones' advance. They sent an urgent plea to their own ancient enemy, Rome, for military help. The Romans, not too excited over giving aid to the Etruscans, sent three envoys to mediate the dispute between the two factions. The dispute was over land. The Clusians refused to part with any portion of their land and the Celts defiantly retorted "to the brave belong all things." One of the envoys, Quintus Fabius, sided with the Clusians and slew a Celtic chieftain. This resulted in the Celts demanding that the Roman Senate deliver up the Fabians and they sent their delegation to Rome to watch the proceedings. Instead of yielding up the Fabians, the Senate gave them the highest honors that could have been bestowed by appointing them consular powers. This was a slap in the face to the Celtics. Enraged, a cry, "to Rome" went out from the blond giants.

A Senone army of thirty thousand giants began their march to Rome. As they neared Rome the Celtics employed psychological warfare by beating their swords rhythmically against their shields, while assailing the air with loud war cries, accompanied by the unceasing blare of innumerable horned helmeted trumpeters. The deafening din frazzled the Roman's nerves to the point that their hair stood on end.

At this point several skirmishes, on the banks of the Allia, between the Romans and Celtics happened. The giants were savvy to the Roman generals strategy and caught them off guard. They pressed their advantage and commenced a slaughter, resulting in the Romans turning tail and running for their lives. Three days after the Battle of Allia, the barbarian hoard ventured into the city of Rome where they encountered only a "deathly hush of silence." The whole population had panicked and fled. The Celts proceeded to ransack the city, then they set it on fire.

The Romans hid in the hills and continued to battle with the Gauls for seven months. Both sides were experiencing famine and incapacitating plagues of fever and dysentery. Finally the Gauls delivered to the Romans a pledge to withdraw from the country upon the payment of a "bushel of gold." The Romans, deeply humiliated, barely raised the ransom by their scales. The devious Celts then produced their scales which weighed heavier than those used by the Romans. When the Romans protested, the Celtic chieftain Brennus insolently threw his heavy broadsword on the scales and further embarrassed the already embarrassed Romans with this harsh threat:

"Woe to the vanquished!" It was the worst humiliation Rome ever suffered in her history, one that would remain a wound that would never completely heal.[3]

About 285 years later, around 102 B.C., the Romans, under command of Marius, avenged their humiliation in massacres at Aquae Sextiae and on the plain of Vercellae. Here the giant tribes of the Teutones and Ambrones were virtually wiped out. The only survivors of the giant Cimbri tribe were those who fled the battlefield and the sixty thousand that were taken prisoner. It is said that well over one hundred thousand Cimbri were killed. This Roman victory was the beginning of the decline of the Celts everywhere and one from which the blond Aryan giants never fully recovered. Other invaders, at later times, pressed them into scattered remnants.

Quayles states in his web site, "that even after crushing the giants with humiliating defeats from which they never recovered, the Romans still were not quite done with them for three of the giants later became their emperor." The first of these was Calius Julius Maximinus, b. 173 A.D. It is said that his hands were so large that he used the bracelet of his wife, Caecilia Pauline, for a thumb-ring. The second giant Roman Emperor was Flavius Jovian, who was acclaimed emperor in early 363 A.D. He stood between eight and one half to nine feet tall. He reigned only seven months, dying on February 16, 364 A.D., the result of accidental asphyxiation from a charcoal fire he had built to warm his room. The third giant Roman Emperor was Maximilian 1 (1493-1519). Maximilian stood above eight feet tall and was endowed with the might and strength of a giant. He claimed direct descent from an ancient race of giant supermen, the Aryan Cimbris.

Maximilian 1 started the theory that the Germanic branch of white blond blue eyed Aryans were direct descendants of a super race. He believed that they were superior in intelligence and in physical strength. He taught that all other races of men were inferior. He contended that their ancestors, the Cimbri, had once made Rome tremble.[4]

Adolph Hitler believed this theory about the Aryan people and sought to re-establish a ruling class of Aryan supermen in Germany. Quayle states in his web site that, "It took World War 11, and the tragic loss of millions of lives to disprove the Aryan myth and check its advance."

Charlemagne (742-814) was the first Germanic ruler to bear the title of Emperor of the West. He ruled over the Carolingian Empire that today would comprise Germany, France, Belgium, the Netherlands, Switzerland, Hungary, most of Italy, and part of Spain. Legends say that he stood eight feet tall and had super strength. It was said that he could straighten four horseshoes joined together, and lift with his right hand a fully equipped fighting-man to the level of his head. Other tales relate that he was so strong that he felled a horse and rider with the blow of his fist and that he would hunt the wild bull single-handed. His army fought an opponent to a standoff. To move things forward Charlemagne proposed an alliance. The leader of the opposing force attended a meeting with two body guards to discuss the proposal. As soon as the man entered the great hall, his body guards were held while Charlemagne stepped forward and promptly split the man from his collarbone to his pelvis with a single blow. The war was over.[5]

Another giant king worthy of mention was Harald Sigurdsson, the Giant Viking King who battled the English around 1066. An English arrow killed the giant in battle when it struck him in the throat.

The Celts and Gauls were of the northern and western European tribes. Josephus states that they were descendants of Gomer, son of Japheth. It appears that the daughters of Gomer were found to be fair by the Nephilim. Anatolia or Asia Minor was the stomping grounds of the Hittites.

From the Egyptian monuments and even their own monuments we learn that the Hittites were a people with yellow skins and "mongoloid" features. They had oblique eyes, protruding upper jaws, and a long nose bridge coming high to the forehead, having a facial countenance described as Mayan. Like the Babylonians, pointed shoes and long conical hats were their trade mark attire.

The first mention of them in the *Bible* is when Abraham purchases the field and the Cave of Machpelah from Ephon the Hittite as a place of burial for his wife, Sarah.[6] Then Esau took his first two wives from the Hittites.[7] The *Bible* also states that the Hittites were the founders of Jerusalem.[8] This means that they probably built and occupied the megalithic city of Ba'albek as it lies between their realm and Jerusalem. Ba'albek is one of the largest stone structures in the world. As far as known one wall, called the Trilithon, is comprised of three blocks of hewn stone that are the largest hewn blocks ever used in construction on this planet. It is an engineering feat that has never been equaled. By conservative estimates the blocks weigh at least 750 tons each or one and a half million pounds. According to ancient Arab legends, the first Ba'albek Ba'al-Astarte temple, which includes the massive stone blocks, was built a short time after the Flood, at the order of King Nimrod, by a "tribe of giants."[9]

The Temple Mount at Jerusalem is built on a foundation of huge ashlar blocks like those of Ba'albek. They used megalithic construction known as cyclopean which was huge odd-shaped polygonal blocks perfectly fitted together. This type of building technique is the most sophisticated known as it is able to withstand powerful earthquakes and last for thousands of years, noting that Turkey and the Mediterranean are earthquake prone. That the Hittites were an ancient tribe of giants is verified by the fact that they occupied and probably built Jericho in 8,000 B.C., as well as the towns of Catal Huyuk, Turkey, and Hebron. Anak's father was Arba, a descendant of the Nephilim, and the original builder of Hebron.[10] They were the builders of "the giant cities of Bashan" also mentioned in the *Bible.* History classically states they appeared in the 16th century B.C., but they are a much older civilization by at least several thousand years.

The Hittites main stomping ground was central Anatolia or what is now called Turkey. Bogazkoy, also called Hattusas or Hatti Land, Turkey, was their capitol city. In their heyday they controlled all of coastal Turkey, present day Lebanon, Cyprus, ancient Crete,

Mycenae, and southward as far as Gaza. They were a warring, sea faring, conquering people, and were regarded as barbarians. In fact, the Hittites waged a war against Egypt from 1300-1200 B.C., that tragically drained both empires. In between wars their primary activity was commerce and trading with all the civilizations of the Mediterranean and ports of Europe. They were also metallurgists and controlled the important tin trade from England. Because they were conquerors of many nations they adopted the social structures of each into their society, their language was a mixture of English, German, Greek, Latin, Persian, and languages of India, called Indo-European language.

The height of their empire was from 1600-1200 B.C., then the Assyrians gained control of Mesopotamia sometime after 1300 B.C. Slowly their empire started to decline and by 717 B.C., their territories were finally conquered by Assyrians and others.[11]

The Amorites were descendants of Ham's son, Canaan, and were called Canaanites in *The Bible.* No doubt the Raphiam, a division of the Nephilim, found Canaan's daughters to be fair and produced a giant race. They lived in close contact with the Sumerians.

Besides being great in statue, they were a handsome people with white skins, blue eyes, reddish or light hair, aquiline noses, and pointed beards. The original seat of the Amorite tribes was the mountain ranges of Taurus of Asia Minor and they were one of the seven nations of Canaan. Later they lived on the southern slopes of the mountains of Judea, which was called "the land of the Amorites." As with other giant tribes, they were war like and frequently warred between themselves. It apparently didn't take much to set them off on a warring prattle. The Philistines and the Amorites are synonymous. *The Bible* says the Raphiam are to have no resurrection in the world to come.[12]

The Hebrews viewed them as warring mountaineers who were evil, idolatrous, and sinful. Their empire reached its peak between the 18th through the 13th century B.C. *The Bible* indicates that Joshua utterly destroyed the giant Amorites, and for the most part he did, but some also managed to escape the general destruction by running off earlier to other countries. Only one word of their language survives—"Shenir", the name they gave to Mt. Hermon.

South America has many megalithic cities with cyclopean and polygonal style walls. They were built by different tribes of giants thousands of years ago, some shortly after the cataclysm and some date later between 7,000 and 3,000 B.C. Later the Incas inhabited some of these ruins and they became Incan cities but they were not the original builders. Often the Incas built on top or added their smaller size structures to the existing ruins. The Incan building technique is easily recognized by its rectangular or square blocks, typically weighing about 200 to 1000 pounds. Beneath their construction one will find the megalithic construction of odd-shaped blocks weighing anywhere from 20 to 200 tons. Some of the most well known megalithic cities in Peru are Cuzco, Kenko, Sacsayhuaman, Tambomachay, Ollantaytambo, and Machu Picchu.

In Bolivia are the famous megalithic ruins of Tiahuanaco that predate the Incas. An ancient legend says that Tiahuanco was built in a single night, after the flood, by unknown giants. The city was probably re- built shortly after the cataclysm at an elevation of 13,500 feet in the Andes, noting that the Andes were thrust up from sea level to present elevations—and still thrusting higher each year—after the cataclysm. The giants chose to rebuild on the same location because they believed that it would be safe ground should another flood occur. The authors believe that after a few hundred years the city was abandoned because of severe earthquake activity. It is interesting to note that some megalithic cities were later built by a newer more sophisticated method of construction that was extremely earthquake proof. This method made keystone cuts into adjoining blocks that were then filled with melted copper or silver, producing walls that were literally stapled together. This type of bracing could withstand most earthquakes.

According to the written *Chronicle of Akakor*, quoted from memory by Tatunca Nara, a descendant and chieftain of the Ugha Mongulala tribe, the Ancient Fathers erected three sacred temple complexes: Salazere on the upper reaches of the Great River, Tiahuanaco on the Great Lake, and Manoa on the high plain in the south. These temple complexes were supposedly the residences of the Former Masters, and were off limits to the tribe.

The *Chronicle of Akakor*, quoted from Memory by Tatunca Nara states:

"And the Gods ruled from Akakor. They ruled over men and the earth. They had ships faster than birds' flight, ships that reached their goal without sails or oars and by night as well as by day. They had magic stones to look into the distance so that they could see cities, rivers, hills, and lakes. Whatever happened on earth or in the sky was reflected in the stones."[13]

If the above information is true, then it may be that Tiahuanaco was built by the Nephilim themselves. The chronicles relate that these Masters built 26 stone cities around Akakor, the largest being Humbaya and Paititi in Bolivia, Emin on the lower reaches of the Great River, and Cadira in the mountains of Venezuela. The chronicles state further, "But all these were completely destroyed in the first great catastrophe thirteen years after the departure of the Gods."

The Masters or Nephilim were described as looking very much like men, with fine features, white skin, bluish black hair, thick beards, and having six fingers on each hand and six toes on each foot. The Former Masters had great knowledge and as if by magic could suspend the heaviest stones, fling lightning, or melt rocks. The first of the 26 cities built by the Masters were Akanis, Akakor, and Akahim. Akanis was built on a narrow isthmus in Mexico.[14]

Tatunca Nara continues in his narration: "After the Former Masters left in the year zero, according to the chief, some sort of global catastrophe occurred. Just before the catastrophe (cataclysm), there was some sort of war between the Gods, (Phaeton's clashes with the planets) something horrible and devastating. After this war and catastrophe, the Ugha Mongulala and the surrounding tribes lapsed into 6,000 years of barbarism."

He continues with the following account, "During this first cataclysm in the year 13 (10,468 B.C.) the course of the rivers were altered and the height of the mountains and the strength of the sun changed. Continents were flooded. The waters of the Great Lake flowed back into the oceans. The Great River was rent by a new mountain range and now it flowed swiftly toward the East. Enormous forests grew on it banks. A humid heat spread over the easterly regions of the empire. In the West, where giant mountains had surged up, people froze in the bitter cold of the high altitudes. The twenty-six cities were destroyed by a tremendous flood. The sacred temple precincts of Salazere, Tiahuanaco, and Manoa lay in ruins, destroyed by the terrible fury of the Gods."

Tatunca Nara states that prior to the cataclysm the continent was, "…still flat and soft like a lamb's back…the Great River still flowed on either side."[15]

It is interesting to note that the year of the cataclysm closely adheres to the year 9,600-500 B.C., given by Allan and Delair in their book, *When the Earth Nearly Died*. The authors believe that Tiahuanaco was partly rebuilt after the cataclysm and later abandoned. There are a few remaining statues in the temple complex that indeed have six fingers and six toes!

On March 3, 1972, a German journalist named Karl Brugger met Tatunca Nara and taped recorded his entire account which he later published in his book, *The Chronicle of Akakor*.[16]

Where there are Nephilim there are also giant offspring. When Ferdinand Magellan discovered Patagonia in 1519, he anchored at Port San Julian, just north of Tierra del Fuego, "The Land of Fire." When they anchored in a natural bay, a giant native appeared on the beach followed by others. Magellan's account states that the heads of his men barely reached the giant's waist, and he was proportionately big. His body was formidably painted all over, especially his face. A stag's horn was drawn upon each cheek, and great red circles round his eyes; his colors were otherwise mostly yellow, only his hair was white.

Magellan named them the Patagons. He noticed that they tied up their short hair, with a cotton lace. They have no fixed habitations, but certain moveable cottages, which they move from place to place; these cottages are covered with the same skin that covers their bodies. A certain sweet root, which they call by the name they give to bread, capar, is a considerable part of their food; what flesh they eat is devoured raw.

Sir Francis Drake encountered these same giants on one of his voyages to Patagonia. Later in another account of a voyage round the world, by Sir Thomas Cavendish, he states, "A wild and rude sort of creatures they were; and, as it seemed, of a gigantic race, the measure of one of their feet being 18 inches in length."

Oliver Van Noort, the first Dutchman that attempted a voyage around the world between 1598 and 1601, apparently encountered these same giants. They killed the grown male giants and took captive four boys and two girls. One of the boys, brought on board

Van Noort's fleet, learned the Dutch language and told them that the inhabitants of the continent near the island from which he had been taken, were divided into different tribes; that three of these tribes were called the Kememtes, Kenekin, and the Karaicks. These three tribes were of common size but that there was another tribe of gigantic stature called the Tiriminen. These giants were 10 to 12 feet high and continually at war with the other tribes. These accounts were all published in *The Gentleman's Magazine*; May 1767, pages 195-197, 238, and 239 by Sylvanus Urban Gent.[17]

Commodore Bryon visited the Magellan Strait in 1764. He and his men had a meeting with five hundred more of these giants. One of Bryon's officers wrote, "...some of them are certainly nine feet, if they do not exceed it. The commodore, who is very near six feet, could but just reach the top of one of their heads, which he attempted, on top-toe; and there were several taller than him on whom the experiment was tried."[18]

The native-born Mexican scholar, historian, and Catholic priest Fernando de Alva Ixtlilxochitl (1578-1650 A.D.) is considered by many to be the most prolific writer on the early history of Mexico. Essentially he can be compared to as the Josephus of Mexican history. He was born of royalty, being a descendant of both the last king of Texcoco and the next-to the last Emperor of Mexico. He was a descendant of the Chichimeca people.

In regards to the sources for his history of Mexico, Ixtlilxochitl wrote the following:

"...of a truth I have the ancient histories in my hand, and I know the language of the natives, because I was raised with them, and I know all of the old men and the principals of this land.... It has cost me hard study and work, always seeking the truth on everything I have written...."

His accounts are taken from many manuscripts that were circulated in the year 1600 A.D. Few writers have enjoyed the fame and reputation that he has yet his numerous works are largely unknown. When Lord Kingsborough of England published nine volumes of work entitled *Antiquities of Mexico*, he included the writings of Ixtlilxochitl in Spanish. This work was published in 1832-1848, but because of the extensive cost of publication it was never widely circulated. In 1965, Alfredo Chavero edited and annotated the works of Ixtlilxochitl. It was republished in two volumes of approximately 500 pages each with a preface by Lic. J.Ignacion Davila Garibi. Chavero called the books *Obras Historicas de Don Fernando de Alva Ixtlilxochitl.*[19]

In the accounts of Ixtlilxochitl, he relates that the first people, the Quinametzin, came from the great Tartary, and were part of those who came from the division of Babel to this land. They came from the great tower at the time of the confusion of languages, wandered for 104 years before they settled at Huehue Tlapallan, which became their capital city and means "Ancient Place of the Red."---Huehue meaning "ancient," and Tlapallan meaning "place of the red." Their king, Chichimecatl, traveled with them crossing a large part of the world arriving at this land which they considered to be good. Afterwards they settled in the northern part of the land. Thereafter they called themselves Chichimecas, named after their first king. It was customary wherever the Chichimecatl settled, whether it be a large city or small town, to name it after the first king or leader who possessed the land. Among the Chichimecas were divisions or branches. Some were more civilized and some were more barbaric but they all are descended from the same forefathers; and as it has been said, they came from the Occidental areas.

In 1554, in the ancient Guatemalan town of Totonicapan, the prominent native nobles of the town wrote a document called the *Titulo de Totonicapan,* that told the origin of their ancestors and is similar to Ixtilxochitl's above writing. The Quiches are the Indian people of Guatemala who are the descendants of the Maya. The authors of the *Titulo de Totonicapan* write:

"The three wise men, the Nahuales, the chiefs and leaders of three great peoples and of others who joined them, called U Mamae—the ancients—extending their sight over the four parts of the world and over all that is beneath the sky, and finding no obstacle, came from the other part of the ocean, from where the sun rises, from a place called, in Mayan, Pa Tulan, Pa Civan.

The principal chiefs were four. Together these tribes came from the other part of the sea, from the East, from Pa Tulan, Pa Civan, bordering on Babylonia." These, then, were the three nations of Quiches, and they came from where the sun rises, descendants of Israel, of the same language and same customs.

When they left Pa Tulan, Pa Civan, the first leader was Balam-Quitze, by unanimous vote, and then the great father Nacxit {God} gave them a present called Giron-Gagal. When they arrived at the edge of the sea, Balam-Quitze touched it—the sacred director—with his staff and at once a passage opened, which then closed up again, for thus the great God wished it to be done, because they were the sons of Abraham and Jacob."[20]

It is interesting to note that the sacred ball or director is also mentioned in the *Book of Mormon* in *Alma* 38:38 as follows: "And now, my son, I have somewhat to say concerning the thing which our fathers call a ball or director—or our fathers called it Liahona, which is being interpreted, a compass; and the Lord prepared it." From this it would appear that these records are speaking about the same peoples. Ixtlilxochitl then gives a surprising account of giants who were called Quinametintzoculithicxime. It is believed that this tribe of giants was one of the branches of the Quinametzin/ Chichimecas.

From the above accounts, it appears that the Nephilim also found the daughters of the first settlers, the Quinametzin, to be fair. Their giant offspring became a race called the Quinametintzoculith icxime. The ancient Toltec record keepers referred to the giants as Quinametzin—also known as Quinames—and as they had a record of the history of the Quinametzin, they learned that they had many wars and dissensions among themselves. Joe Allen states in his book, *Exploring The Lands of The Book of Mormon*, "that the giants were destroyed, and their civilization came to an end as a result of great calamites and punishments from heaven for some grave sins that they had committed." Ixtlilxochitl gives the date of their destruction as 240 B.C. Joe Allen further compares some common elements with Ixtilxochitl's writings and the *Book of Mormon* in the following: (note that Ixtilxochitl's writings predate the *Book of Mormon* by more than 200 years)

1. They both speak of the first civilization coming from the great tower at the time of the confusion of tongues.

2. They both speak of a white god who was born of a virgin and who ascended to heaven after teaching his people.

3. They both record the date of a great destruction occurring in the first month of the 34th year, or at the death of Christ.

4. They both use the same terminology in describing the manner in which cities were named.

5. They both speak of three distinct civilizations that predate the coming of Christ.

6. They both record the destruction of the first civilization that predates the coming of Christ, who lived in the northern lands, or the Land Northward.

7. They both speak of a nation whose principal area meant "land of abundance" or "bountiful."

Comparison Chart

Ixtilxochitl's Writings	*Book of Mormon*
QuinametzinQuinametintzoculihicxime/Quinames	Jaredites
Chichimecas/Ulmecas/Olmecs	Jaredites
Xicalancas	Mulekites
Toltecas	Nephites

The Toltecas were the record keepers, having a written language. They were wise men of the arts and sciences. The same is said of the Nephites in the *Book of Mormon.* If the Toltecs and the Nephites are one and the same, then the *Book of Mormon* is an ancient Toltec record. It is also interesting to note that the Maya have legends surrounding their lost golden books and by comparison the fact is that the ancient Toltec record known as the *Book of Mormon* was also written on gold plates.

Joseph Smith, the Mormon prophet, who translated the ancient Toltec/Nephite records and published them as the *Book of Mormon* in 1830, once expressed an opinion as to where the *Book of Mormon* lands were located. In 1841, John Lloyd Stephens published a book about his travels in Central America, Chiapas, and Yucatan. Expressing his opinion Joseph Smith said,

"Mr. Stephens' great development of antiquities are made bare to the eyes of all the people by reading the history of the

Nephites in the *Book of Mormon*. **They lived about the narrow neck of land, which now embraces Central America, with the cities that can be found."**[21]

From these comparisons it appears that the Ulmecas were the Olmecs and they were a branch that descended from the Quinametzin. The Mayans were also a branch that descended from the Quinametzin. Since the Quinametzin came from the Oriental section of the world, they and their descendants were oriental in appearance and brought their oriental culture to the new world. (the yin and yang symbols) The Olmecs have been described as being a black race but if one is to visit the Veracruz area of Mexico today, you will see many faces with similar features but they are not black but rather Polynesian looking with brown skins. They too have been described as a large people, not tall but large. Their descendants in the Veracruz area today have the same body build. The first settlers were probably a mixture of several peoples from the Tartary section of the Orient thus the different looks.

The Quinametzin or Jaredites left the Tartary section of the Orient about 2800 B.C., and arrived in Mesoamerica around 2700 B.C. The *Book of Ether* found in the *Book of Mormon* gives their account of how a colony of twenty-four families, led by the hand of God, arrived on this continent coming in eight ship like barges that crossed the great waters in 344 days.[22] The *Book of Mormon* has the abridged record of the family lineage of just one of these-Jared. Nothing else is mentioned of the other twenty-three family histories. These other families were the different branches of the original

Quinametzin settlers that Ixtlilxochitl spoke of in his account of the first people.

However, the *Popol Vuh* gives the record and history of most of the other tribes, who arrived from the East whose names are found in the *Titulo de Los Senores de Totonicapan.* Ixtilxochitl's writings tell of a second migration of people called the Xicalancas who came in ships or boats from the east to the land of Potochan (now Veracruz), and from there they began to populate the land. Upon arrival they soon found some of the giants who had escaped the destruction and extermination of the calamity, called the second age. These giants because of their size and strength oppressed and enslaved their new neighbors. To free themselves of the giants, the new settlers invited them to a solemn feast. After the giants became full and intoxicated, they were killed and destroyed with their own weapons.

In comparison, the *Book of Mormon* calls these people the Mulekites. Mulek was the son of Zedekiah/Mattaniah. It is speculated that Mulek was a young baby who was disguised as a daughter, or perhaps still in his mother's womb which allowed him to escape the wrath of death instituted on each of the sons of Zedekiah by King Nebuchadnezzar at 586 B.C. In time he and his peoples, having multiplied greatly, were also led by God to Mesoamerica. They came on ships, perhaps on Phoenician ships and landed in the Land Northward, which was the area of the heartland of the Jaredites. Mulek was a Jew of the tribe of Judah.

The Mesoamerican document called *The Lords of Totonicapan*, written in 1554, may be referring to the Mulekites in the following:

".... they came from where the sun rises, descendants of Israel, of the same language and the same customs."[23]

There is an oral tradition that still prevails in Veracruz, which relates that many in the area are descendants of the House of Israel.

Ixtilxochitl's writings tell of the third and last migration to early Mexico, the Toltecas. By all accounts they are described as a bearded white race, "high of stature", who came from the east on ships. He also states that as late as the tenth century A.D., there were white, blond children born to Tulteca descendants.[24] Many historians believe that they were Phoenicians but the *Book of Mormon* by comparison states that they were of the House of Israel originally living in Jerusalem and were descendants of the tribe of Joseph.[25] Lehi and his colony left Jerusalem about 601/600 B.C., and after much wandering in the wilderness they were directed by God to build a ship and cross the waters to the new "promised land." The *Book of Mormon* calls them Nephites and later a splinter branch became known as the Lamanites. Joe Allen, in his book, identifies the Lamanites as the Classic Mayas (200-900 A.D. time period).

The writings of Ixtlilxochitl tell of the Spaniards traveling along the coastal areas, such as the lands of the Chicoranos and the Duharezases, and finding men in those parts who are eleven and twelve hands in height, and they were told that there were others even taller who resided inland that were thirty hands tall.

The megalithic cities of Monte Alban, and Mitla, Mexico, were probably built by the giant Quinametzin and the large Olmecs. As with other ancient sites newer construction has been added on top the original structure.

Besides building megalithic cities the Nephilim and their giant offspring apparently were involved in digging tunnels through the mountain ranges of the world that ran for hundreds, even thousands of miles, and all interconnected. They used these passages for traveling, protection, and storage of manuscripts and treasure. There are legends of these tunnel systems in Tibet, India, South America, America, parts of Central Asia, Britain, and Europe. All say the giants built them. Do they exist? In 1972, after a big earthquake in Lima Peru, a tunnel system was discovered beneath that coastal city. On exploration of the tunnel system the explorers were amazed to find that large parts of the city were undercut by a maize of tunnels, all leading into the mountains. These tunnels couldn't be explored to their terminal points because they had collapsed during the course of the centuries. A similar tunnel system was discovered beneath Cuzco in 1923 that went in the direction toward Lima. In Guatemala, tunnels have been discovered that run for as much as 30 miles in one direction that linked Mayan cities in a vast underground network.

Another set of similar tunnels may exist in Arizona. When the American army was in hot pursuit of Geronimo around Arizona; he and his braves would ride into box canyons and then literally disappear leaving the U. S. Army totally mystified. Later, Geronimo and his braves would suddenly turn up, hundred of miles away in Mexico! This event happened many times. Was Geronimo using the passage ways of these ancient tunnels to escape? Certain members of the Navaho tribe apparently know about these tunnels, but keep them secret.[26]

The monolithic sarcophagus of Pacal, the king of Palenque, is a large one. Pacal has been said to have been nine feet tall or more.

Pacal passing his throne onto his son. Notice the difference in height between the giant king and his son.

The highly carved lid of his sarcophagus is nearly thirteen feet long. This would indicate that he was exceptionally tall in stature for a Mayan, who as a people are short. The Tablet of the Foliated Cross and the Tablet of the Cross at Palenque, both confirm that Pacal was at least nine feet tall. In these carved tablets Pacal is shown opposite his son, Kan-Bahlum, and in both carvings the differences in height in obvious. He ascended to the throne through his mother, who was a Mayan ruler for a short period of time as was his great-grandmother. No mention is made of his father but he claimed divine kingship. Could he have been a son of the Nephilim or a descendant of one of the survivors of the giant Quinametzin in the 240 B.C. destruction?

There were giants in Africa. The following is taken from Steve Quayles web site:

"In 1936, two French archaeologists, Lebeuf and Griaule, led an expedition to Chad in North Central Africa. As they crossed the plains they saw some areas covered with small mounds. They also found large number of these mounds around Fort Lamy and Goulfeil. Deciding to investigate, they dug up several egg-shaped funeral jars that contained the remains of a gigantic race, along with pieces of their jewelry and their works of art. These giants, according to the natives, were called the Saos."

"Scholars who traced their history say they came from Kheiber, located north of Mecca, to Bilma, which is situated about three hundred miles north of Lake Chad. A people with a "well-developed religion and culture," they grew in numbers and founded communities at Fort Lamy, Mahaya, Midigue, and Goulfeil. They lived in peace in their new land until the close of the ninth century when the Moslems made wars against them, intending to force their acceptance of the Islamic faith. The Saos giants who converted to the faith lived to become servants of the Arabs. But those who steadfastly refused to convert were eventually wiped out. By the end of the sixteenth century not many Saos remained."

Are the Watusi descendants of the giants? Steve Quayle states in his web site the following, "The Watusi are black, but they are not Negroes. Many grow to heights of seven feet or more. Anthropologists are at a loss to explain the Watusi's tallness. One possible explanation is that they are offspring of the giants who fled before Joshua's legions and escaped to Africa, but, after many centuries of interbreeding with the aborigines, have been greatly reduced in bulk and might."

Marco Polo tells of running into a gigantic people in Zanzibar, an island on the African coast. He wrote the following account:

"Zanzibar is a very large and important island. It has a 2,000 mile coastline. All the people are idolaters, they have a king and a language of their own and pay tribute to no one. The men are large and fat, although they are not tall in proportion to their bulk. They are strong limbed and as hefty as giants. They are so strong that they can carry as many as four men, they eat enough for five. They are quite black and go about completely naked but for a loincloth. Their hair is so curly that they can only comb it when it is wet. They have wide mouths and turned-up noses...."

Many islands upon the sea have megalithic structures that were built by the giants at one time or another. More of this will be discussed in the Lost Civilizations chapters.

In the Kiribati Islands located in the Pacific Ocean, there is a legend of two giants, apparently brothers, who came to the island of Nauru, from the sky. They were twice as large as normal humans and could lift enormous stones and perform other feats that required great strength. One of the giants was named Nauru, hence the name of the island. Because the islanders were afraid of them, they devised a plan to get the giants drunk on palm wine and then kill them. After they killed the giants, they were buried in pits and covered with stones. The local natives have shown the two gravesites to several investigating groups. Nearby are several upright monoliths, apparently compass stones. Two of the stones were made of granite which is not found on the Kiribatis. Besides the gravesites there are many fossilized huge footprints that can be clearly seen in the volcanic rock through out the islands. The main spot that these footprints can be found is in the village of Banreaba. Most of these footprints are three feet long and have six toes on each foot.[27]

On Pohnpei Island, formerly called Ponape, in Micronesia are the gigantic ruins of Nan Modal. No one knows who, when, or how the city was built, but the Japanese reportedly did discover very large human bones there that indicated the previous inhabitants were seven feet tall or more. The local natives have a legend that the stones magically flew through the air.[28]

In Australia, the Aborigines have a tradition that there once lived a race of giants in their land who stood from twelve to fifteen feet tall. Evidence of a past presence is preserved in fossilized enormous foot tracks in central Queensland near Bathurst as well as on the Blue Mountains of New South Wales. In the 1960's, near

Bathurst, enormous hand-axes, clubs, adzes, knives, and other giant tools were excavated that could only be used by people of immense strength and height; beings over twice the height of modern man. Even today it is said that there is a race of seven feet tall giant Aborigines living in the remote Simpson Desert in the middle of the continent. Childers states in his book, *Lost Cities of Ancient Lemuria & The Pacific*, the following:

"Giant stone chairs can be found in the central desert and all over the country. There are stone circles and Stonehenges along the northern New South Wales coast on the plateau behind Coff's Harbor. In the Simpson Desert are hundreds of standing stones and megaliths, some of which stand thirty feet high. They appear to have come from miles away! Australia is full of this stuff."

Many giant human footprints contemporary with dinosaurs have been found in Dinosaur Park near Glen Rose, Texas. One particular footprint was determined to be that of a woman estimated to be about 10 feet tall and weighing 1,000 pounds. These estimates were determined through cross-sectional cuts and compression studies. In the same area, some giant human footprints have been found inside the tracks of dinosaurs. Glen Rose has a museum where plaster castings of the original tracks are on display. Scientist say these findings contradict modern evolution theories because mankind was not to evolve for yet another 75 to 100 million years, but these giant foot prints were Nephilim giants, not human.

The stories and accounts of giants covered in this chapter are just a few of the many hundreds. What has been written about them in this chapter barely scratches the surface. The most important thing that the authors have tried to convey is that a race of giants once occupied this planet both before and after the cataclysm. This chapter has dealt with the second irruption of Nephilim that occurred shortly after the cataclysm. The Lost Civilization chapters will tell of the first irruptions of the Nephilim prior to the cataclysm and of events leading up to their destruction.

For some unknown reason history has largely ignored the worldwide existence of these giants. By doing so a huge gap of several thousand years remains historically empty of their influence and presence upon this planet. If history would recognize their existence it would go far to solve many problems connected with Anthropology.

The irruption of the Nephilim was the first attempt to stop the seed of the woman from being born; and was directed against the whole human race because it wasn't made known to Satan which lineage of Adam the "seed of the woman" would be born. In time God made known through his prophets the royal lineage through which the promised "seed of the woman" should come.

So when Abraham was called by God, then Satan attacked him and his seed, perceiving him to be the first of the royal lineage.

When David became king, then Satan assailed the royal line.

And when "the Seed of the woman" was born Satan caused Herod to seek the young child's life.[29] Later, Satan tempted Jesus to "Cast Thyself down."[30] There was another attempt to cast him down and destroy him. Again Satan fought for eternal dominion of this world by asking Jesus to "fall down and worship me."[31] Satan sought to destroy Jesus as he slept on a ship by causing a great storm to arise in an attempt to sink the vessel.[32]

All these attempts failed to stop the "Seed of the woman" in fulfilling his destiny to redeem mankind and restoring to Adam and his righteous seed, dominion of the future world to come. Until that time, Satan has dominion of this planet, but his time is short.

According to the biblical scriptures the fallen angels who sinned before the flood in the spiritual creation are now chained in Tartarus and they remain imprisoned awaiting the judgment of "the great Day."[33] They are unable to do further harm to mankind. However, the Nephilim who came to the temporal creation, both before and after the cataclysm and continued after the arrival of Noah, who caused the genetic defects in their giant offspring remain unrestrained until this day as far as we know. Similar unions could occur again as the rise of demonic activity in the age we live in is anticipated by the *Bible*.

Isaiah 24: 21-22 assures us that all remaining fallen angels will be judged and removed in a future day from all their influence and power over both nature and the affairs of mankind.

"And it shall come to pass in that day, that the Lord shall punish the [angelic] host of the high ones that are on high, and the [evil] kings of the earth upon the earth.

And they shall be gathered together, as prisoners are gathered in the pit, and shall be shut up in the prison, and after many days shall they be visited."

Ephesians 6:11-12 warns us in the following verses of the power of Lucifer, the prince of darkness:

"Put on the whole armour of God, that ye may be able to stand against the wiles of the devil. For we wrestle not against flesh and blood, but against principalities, against powers, against the rulers of the-darkness of this world, against spiritual wickedness in high places."

End Notes

[1] *(2 Samuel 21:22 and 1 Chronicles 20:8)*

[2] (*Job* XXXVI, 5). (*Zohar Vol. V*, Sperling & Simon, The Soncino Press, 1984, p. 229)

[3] (www.stevequayle.com/Giants) (viewed March 2003)

[4] (www.stevequayle.com/Giants)

[5] (www.stevequayle.com/Giants) (The World Book Encyclopedia)

[6] *(Genesis 15:20; 23:3-18)*

[7] *(Genesis 26:34; 36:2)*

[8] *(Numbers 13:29-30)*

[9] *Lost Cities of Atlantis, Ancient Europe & The Mediterranean,* ©1996 by David Hatcher Childress. Published by Adventures Unlimited Press, Kempton, ILL., pp.31, 33,

[10] (*Genesis* 35:27; Joshua 15:13, 21:11)

[11] *Lost Cities of Atlantis, Ancient Europe & The Mediterranean,* ©1996 by David Hatcher Childress. Published by Adventures Unlimited Press, Kempton, ILL., pp.58-61.)

[12] *(Isaiah* 26:14)

[13] *Lost Cities & Ancient Mysteries of South America,* ©1986 by David Hatcher Childress. Published by Adventures Unlimited Press, Kempton, ILL. All rights reserved., p. 247)

[14] (et al, p. 245)

[15] (et al, p. 248)

[16] (*The Chronicle of Akakor*, Karl Brugger, 1977, Delacorte Press, NYC)

[17] (www.stevequayl.com/Giants/articles/Gentlemans.Magazine.html viewed March 2003)

[18] *Lost Cities & Ancient Mysteries of South America,* ©1986 by David Hatcher Childress. Published by Adventures Unlimited Press, Kempton, ILL. All rights reserved., p. 197

[19] *Exploring the Lands of The Book of Mormon,* ©1989 by Joseph L. Allen, S.A. Publishers, Inc., Orem, Utah. pp.137-139)

[20] (*New Evidences of Christ in Ancient America*, Yorgason, Warren,& Brown, 1999, *Book of Mormon* Research Foundation, pp.42, 45, 47, 74 as cited in Recinos and Goetz, 1953, pp.194, 169-70)

[21] *(Times and Seasons*, 3:915, emphasis added). (*New Evidences of Christ in Ancient America*, Blaine M. Yorgason, Bruce W. Warren, Harold Brown, *Book of Mormon* Research Foundation, 1999, p. 118)

[22] *(Ether* 6:14-16)

[23] (Recinos and Goetz 1953:170)

[24] (*New Evidences of Christ in America*, p. 75)

[25] (2 Nephi 25:5)

[26] (*Lost Cities & Ancient Mysteries of South America,* ©1986 by David Hatcher Childress. Published by Adventures Unlimited Press, Kempton, ILL. All rights reserved., pp.64, 67.)

[27] (*Lost Cities of Ancient Lemuria* & The Pacific, David Hatcher Childress, Adventures Unlimited Press, Stelle, Illinois, 1988, pp.191, 192,193)

[28] (Same as above, p. 211, 213, 222)

[29] (*Matthew* 2)

[30] (*St. Matthew* 4:6) At Nazareth, again (*Luke* 4)

[31] (*Matthew* 4:9)

[32] *(St. Mark* 4:35-41)

[33] *(Jude* 6)

E. J. Clark & B. Alexander Agnew, PhD

The Lost Civilization of Mu

According to legends the first civilization arose, on the temporal earth, about 78,000 years ago on a large landmass that was divided into three large islands. Australia was the southern most island mass. Supposedly, the Pacific Islands are the remaining mountain peaks of the two other lost island lands. The great Lemurian empire may have comprised the Fiji Islands, Hawaii, Easter Island, and some of the Los Angeles area.

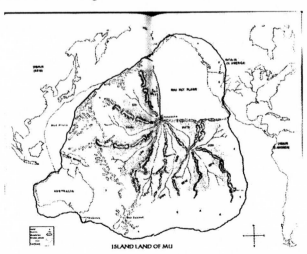

ISLAND LAND OF MU

379

Nostradamus, in quatrain 4-15, speaks of a land which he refers to as the "eye of the sea." In quatrain 1-50 he refers to a land which he designates as "the aquatic triplicity." Australia, New Guinea and Tasmania are all geologically part of the one piece of continental shelf and, could also be the triple island complex. The region was a tropical paradise, described much like the Garden of Eden. It was called Mu or Lemuria, but sometimes referred to as Pan, Rutas, Hiva, and Pacifica.

In 1860, a group of British geologists noted similarities between the strata of the Permian age in India, South Africa, Australia, and South America. Identical fossils of land plants and land animals were also found in these strata from the different continents. Geologists postulated the existence of former land bridges and even continents that had long ago sunk beneath the oceans, to explain their findings in the similarity of the strata.

Ernest Heinrich Haekel used the hypothetical land bridge in his theory to explain the distribution of Lemurs in Africa, India, Madagascar, and the Malaysian Peninsula. He theorized that a land bridge once existed long enough above water to allow Lemurs to spread into these regions. This "land-bridge" was named *"Lemuria"* by an English biologist, Philip L. Scalter, because of the theorized association with Lemurs.

Later, better theories explained the similarities of the strata and fossils. It became clear that land bridges and sunken continents never really existed but the name *"Lemuria"* has stuck and been forever applied as another name for Mu or more correctly, Muror.

The well- known modern occultist, Madame Blatvatsky, reincarnated Lemuria as a lost continent. She claimed to have learned of the existence of Lemuria in a book titled, *The Book of Dzyan*, which was supposed to have been composed in Atlantis and shown to her by the Mahatmas. Sanskrit legends tell of a former continent named Rutus that sank beneath the sea long ago in the Indian Ocean, that obviously she became confused with, however she did give recognition to Philip Scalter for inventing the name of *Lemuria*.

Occultists have developed all kinds of tales about Lemurians, who were like thought projections and not physically solid. They came to earth to experience the physical vibration and were androgynous, neither male nor female. Some chose water, some the plant kingdom, some the mineral kingdom, and some the animal kingdom to project into and experience that reality. They were interdimensional souls that used the rainbow spectrum as a window to move between realities.

In the latter part of the 19th century when all manner of occult and theosophical speculation was rampant, along came the Anglo-American explorer, James Churchward, who later called himself 'Colonel', published during the 1920's and 1930's five main volumes of the Mu series. These popular books were:

1. The Lost Continent of Mu

2. The Children of Mu

3. The Sacred Symbols of Mu

4. The Cosmic Forces of Mu

5. The Second Book of the Cosmic Forces of Mu

It is by studying various ancient texts and tablets that explorer James Churchward claims to have discovered the existence of a long lost continent with an advanced civilization, called *The Empire of the Sun*. This empire, over time, gradually sunk below the Pacific Ocean in stages, starting about 26,000 years ago, following a cataclysmic event. The following is quoted from his first book of the Mu series:

"The Garden of Eden was not in Asia but on a now sunken continent in the Pacific Ocean. The biblical story of Creation came first not from the peoples of the Nile or the Euphrates Valley but from this now-submerged continent, Mu—the Motherland of Man.

From various records it would seem that this continent consisted of three separate lands, divided from each other by narrow seas or channels, but where or how these divisions were made by nature there is nothing to show except possibly, an Egyptian hieroglyph which represents three long narrow lands running east to west.... The land of Mu was half of the Pacific Ocean."[1]

Churchward describes the land of Mu as follows:

"It was a beautiful tropical country with vast plains. The valleys and plains were covered with rich grazing grasses and tilled fields, while the low rolling hill lands were shaded by luxuriant growths of tropical vegetation. No mountains or mountain ranges stretched themselves through this earthly paradise, for mountains had not yet been forced up from the bowels of the earth.

The great rich land was intersected and watered by many broad slow running streams and rivers, which wound their sinuous ways in fantastic curves and bends around the wooded hills and through the fertile plains. Luxuriant vegetation covered the whole land with soft, pleasing, restful mantle of green. Bright and fragrant flowers on tree and shrub added coloring and finish to the landscape. Tall fronded palms fringed the ocean's shores and lined the banks of the rivers for many a mile inland. Great feathery ferns spread their long arms out from the river banks. In valley places where the land was low, the rivers broadened out into shallow lakes, around whose shores myriads of sacred lotus flowers dotted the glistening surface of the water, like vari-colored jewels in settings of emerald green.

Over the cool rivers, gaudy-winged butterflies hovered in the shade of the trees, rising and falling in fairy like movements, as if better to view their painted beauty in nature's mirror. Darting hither and thither from flower to flower, hummingbirds made their short flights, glistening like living jewels in the rays of the sun. Feathered songsters in bush and tree vied with each other in their sweet lays.

The chirpings of lively crickets filled the air, while above all other sounds came those of the locust as he industriously "ground his scissors," telling the whole world all was well with him. Roaming through the primeval forests were herds of "mighty mastodons and elephants" flapping their big ears to drive off annoying insects.

The great continent was teeming with gay and happy life over which 64,000,000 human beings reigned supreme. All this life was rejoicing in its luxuriant home."[2]

Churchward continues in *The Lost Continent of Mu*:

"The dominant race of the lands of Mu was a white race, exceedingly handsome people, with clear white or olive skins, large soft dark eyes, and straight black hair. Besides this white race, there were other races, people with yellow, brown, or black skins.... These ancient inhabitants of Mu were great navigators and sailors who took their ships over the world "from the eastern to the western oceans and from the northern to the southern seas...." And the land of Mu was the Mother and the center of the earth's civilization, learning, trade, and commerce; all other countries throughout the world were her colonies or colonial empires."

We continue on with another excerpt taken from *The Lost Continent of Mu*:

"They that were nine upon nine saw the great High Temple of Mu built. Each stone, each inner part was of them. Then they rested from their work and waited for the great Ra of Mu to come in his golden ship and anchor atop the temple.

Lo! The High Ra Mu, religious leader of the Sun called forth unto nine the Priest. And then he called unto nine the Priestess.... These of the Order of the Phoenix were to each send their nine outward in every direction of the hub. They journeyed to North America, South America, the moon nation, the nation of the Nile. Like a giant spider web they covered the land and great temples were built, and in them were placed light and knowledge.

The sacred writing of the Motherland which were carried by the Naacals, (Holy Brothers) to Mu's colonies throughout the world

70,000 years or more ago…is the oldest written information about the origin of Freemasonry. The extreme age of the brotherhood is not only attested to by the Sacred Writings, but by various Oriental writings, inscriptions, and prehistoric Mexican stone tablets which are, as shown by some of them, over 12,000 years old."

Quoting from the Lhasa Record of Tibet, Churchward continues:

"When the star of Bal (Baal) fell on the place where now is only the sky and the sea, the seven cities with their golden gates and transparent temples, quivered and shook like the leaves in a storm; and, behold, a flood of fire and smoke arose from the palaces. Agonies and cries of the multitude filled the air. They sought refuge in their temples and citadels.

Twice Mu jumped from her foundations; it was then sacrificed by fire. It burst while being shaken up and down violently by earthquakes…. the land was rended and torn to pieces…. quivering like the leaves of a tree in a storm…. rising and falling like the waves in the ocean…. and during the night it went down…."[3]

While serving as an officer in the British Army in India, in 1868, Colonel Churchward befriended a high priest of an Indian temple. The old priest found him to be a seeker of ancient knowledge and eventually allowed Churchward to access some ancient tablets that had been preserved for nearly fifty thousand years. These tablets were only a few fragments of what had once been a vast collection that were rescued from one of the old seven Rishi cities which were the centers of learning in the Rama Empire in ancient India. Over the full course of two years, the old priest taught Churchward how to decipher the bass-relief characters upon the clay tablets. These tablets had been written either in Burma or perhaps even in Mu itself by the ancient Naacal priests. After many months of study, Churchward discovered that the tablets described in detail the creation of the Earth and of the appearance of man on a continent in the Pacific called **the land of Mu.** Supposedly he learned as well, from the texts, of an ancient migration of Naacal priests who left Mu to establish a colony in India. When he fully realized the enormous significance of his discovery, Churchward set off to Burma, in the hope of finding more of the tablets. There the Buddhist priests told him to go back to India because their tablets with all their priceless information had been stolen and carted off to India by thieves. The angry priest then spat at Churchward's feet and quickly walked away.[4]

Undaunted, Churchward became obsessed with finding proof of the amazing story. He studied all the writings of the ancient civilizations of the old world and compared them with the tablets on Mu he had deciphered. He discovered that they were all preceded by the Civilization of Mu. For the next forty-five to fifty years of his life, he became determined to find proof of the actual existence of Mu.

Churchward offers proof firstly via the Naacal tablets. Your authors note that apparently no one else has ever seen these tablets to document or verify that they even exist. He mentions records written in Mayan, Egypt and India, which recount the destruction of

Mu: *"when the earth's crust was broken up by earthquakes and then sank into a fiery abyss. Then the waters of the Pacific rolled in over her, leaving only water where a mighty civilization had existed."*

He makes references to the **Mayan Troano Manuscript** and to yet another Mayan book, as old as the Troano Manuscript, called the *Codex Cortesianus*, which according to him, makes mention of **The Land of Mu.** Again your authors note that these manuscripts translate differently and that no mention of Mu is found anywhere in these writings. The following is taken from an online educational library:

"The *Madrid Codex* was originally identified as two separate manuscripts of unequal lengths called the *Manuscrit Troano* and the *Codex Cortesianus*. The two parts were identified as elements of one codex and were reunited in 1892. The original *Madrid Codex* is now housed in the Museo de America in Madrid and is sometimes referred to as the *Codex Tro-Cortesianus*. The leaves displayed show a portion of the almanac section used by priests to perform divination rites relating to daily activities such as hunting, weaving, and agriculture. The lower half of the panel is composed of glyphs of the 20 named days which, as in the Aztec calendar, cycle 13 times through the 260-day Sacred Year.

Sky serpents that send the rain and speak of thunder are shown weaving around the rows of glyphs. Circles with a cross-hatch design drawn on the bodies of the serpents symbolize *Chicchan*, the 5th of the 20 named days perhaps associating that named day with rain gods or rain ceremonies. In leaf 16 a death god is depicted wearing his characteristic "death eyes" in a collar, as well as at his wrists and ankles. The glyph just above and to the right of his head is the symbol for both death and the sixth named day, *Cimi* day-glyph is the third glyph down from the second column from the right. If a priest reading the codex for divination purposes arrived at this symbol, it was considered an omen of death or illness.

Just above the *Cimi* glyph is the symbol for *imix* (the 1st of the 20 named days), an auspicious sign of abundance when reading portents. Above and to the right of the god, resting on the red horizontal line is the maize god. He holds in his hand the sign of *Kan* (symbolizing both maize and the 4th day), on top of which appears the *Imix* glyph. In leaf 17 two rain gods, called *Chacs* (one is upside down), are above the rows of glyphs. This name is the same Mayan name of lay-priests called *Chacs*, under special priests called *Nacons* and regular priests called *Chilans*. The Mayan civilization reached its zenith by 200 A.D., and inexplicable declined about 600 A.D."[5]

From the above translation it appears that Churchward had an over active imagination as the above text contains primarily astrological and calendrical data. He fails to mention by name what the other records from Egypt, Greece, Central America, and Mexico were. He mentions whilst at **Uxmal**, in the Yucatan, there is an inscription upon an ancient ruined pyramid which commemorates *"The lands of The West, whence we came"* and a pyramid south west of Mexico City, according to its inscriptions, *"in memory of the destruction of these lands of the West"*.

Again, one of your authors has visited these sites and as far as I know these inscriptions do not exist. Churchward cites the Lhasa Record as speaking of seven principal cities, which were the seats of religion, science and learning in Mu, scattered across the three lands. Once more your authors want to remind the readers that no one has ever seen the supposed Lhasa Record outside of Churchward to document its existence.

So, have your authors debunked the theory of the lost civilization of Mu? Well, maybe not. When you discard all the previous writings on Mu in this chapter, there still remains all the legends that abound in Australia, the South Pacific, Polynesia, and Asia, that this land once existed, that it was the place where man first appeared, and that it was destroyed in a cataclysm. One thing is absolutely obvious and undeniable. There are pyramids, and pyramid-like structures in every corner of the world. They are made of sandstone, adobe, granite, and fashioned out of mounds. They have roughly the same religious purpose, and are the center of all leadership for those cultures. They span in time from about 10,500 B.C., to at least 600 A.D. Their involvement with the origin and eternal life of man is evident in the records of the cultures that built them, or perhaps more importantly inherited them.

Some anthropologists would have you believe that this was a result of traveling explorers who came across a land bridge to the extents of the earth. The choice for the reader is to contemplate the possibility that they are wrong. The Aborigines and many southwest Indian tribes speak of this place as " the dream time" or "the first time" period. Your authors feel that where there is "smoke" there usually is "fire." Is there any surviving evidence today that can validate Churchward's claims that these lands once existed? Perhaps there is.

First, these three island lands were located in a highly unstable region surrounded by the area called the "**Ring of Fire**" in the Pacific Ocean. We prefer to call two of them island lands and not continents, but recognizing that Australia is a continent. Because two of the island lands were probably so close to Australia, Churchward may have assumed they were part of the same continent, thus calling all of this land mass the continent of Mu. This would certainly explain the size of Mu described by Churchward as being 5,000 miles from east to west and over 3,000 miles from north to south. He writes *"The continent consisted of three areas of land, divided from each other by narrow channels or seas."* Also it is not unreasonable to believe that perhaps Australia, at one time, had connecting land- masses, separated only by narrow channels.

Supposedly about 50,700 B.C., there was a cataclysm that sunk a portion of the lowlands of a land mass that bordered southern California. Then 24,000 B.C., another island sank beneath the ocean after volcanic eruptions and severe earthquakes. Finally, in 9500 B.C., a final cataclysm nearly destroyed the earth and buried the last of this immense land mass beneath the Pacific ocean. Since this is an earthquake prone region where tectonic plates rub against each other setting off volcanic activity, this certainly adds some credence to Churchward's story.

Was the continent of Mu the cradle of civilization where man first appeared? It certainly would have been the ideal location with a tropical climate, broad flat plains watered by slow rivers and streams, luxuriant vegetation and teeming with many varieties of animals, birds, and fowl. Churchward compares the land as to the Garden of Eden. Doesn't the maternal mitochondrial DNA study say that man came out of Africa? Yes, it does…but it doesn't say when. Early hominids could have walked out of Africa, and colonized parts of the world before the appearance of modern man.

So whence cometh man?

There is a preponderance of evidence that dispels the modern tradition that man evolved simultaneously in distant corners of the world. In fact, the truth that the modern race of man has a common origin, or at least a single point of contact between two races of man, we maintain as undeniable. There is a common thread of belief in a preexistence, a mortal growth or probationary state, and an unshakable believe in eternal life. In every instance the soul received a higher glory as it associates itself with one Creator God while in mortal life. Your authors are now going to present what we found in the ancient writings and scriptures on this very controversial subject.

First remember that there were two creations, the spiritual creation and the temporal creation. Adam, the first man, was created directly out of the spiritual dust of the spiritual earth, by God the father. *The Zohar* or *Book of Splendor*, has many references to the *"upper Adam."* It refers to a *"lower Adam"* as well. The temporal earth was created by God, the son, and other spiritual beings, under the direction of his father. The temporal creation was brought forth differently than that of the spiritual creation. However, the temporal creation is patterned after the spiritual creation. (Readers are referred back to the chapters on the spiritual and temporal creations for details)

The temporal account of the creation is given in *Genesis* 1:25-26. (note: Humans were created *after* the other animals.)

"And God made the beast of the earth after his kind, and cattle after their kind, and every thing that creepeth upon the earth after his kind: and God saw that it was good.

And God said, Let us make man in our image.... So God created man in

His own image."

Genesis 1:27 (The first man and woman were created simultaneously)

"So God created man in his own image, in the image of God created he him; male and female created he them."

The spiritual account of the creation is given in *Genesis* 2:18-19. (Note: The first man was given a name, Adam, and was created *before* the other animals.)

"And the Lord God said, It is not good that the man should be alone; I will make him an help meet for him.

And out of the ground the Lord God formed every beast of the field, and every fowl of the air; and brought them unto Adam to see what he would call them: and whatsoever Adam called every living creature, that was the name thereof."

Genesis 2:20-22 (The man, Adam, was created first, then the animals, and then the woman, Eve, was created from Adam's rib.)

"And Adam gave names to all cattle, and to the fowl of the air, and to every beast of the field; but for Adam there was not found an help meet for him.

And the Lord God caused a deep sleep to fall upon Adam and he slept: and he took one of his ribs, and closed up the flesh instead thereof;

And the rib, which the Lord God had taken from man, made he a woman, and brought her unto the man."

In a brief review, it is important that we grasp clearly one idea. Originally man was **created** in the image of God, the father, who was a spiritual being. The first man, Adam (the upper Adam), was a spiritual entity or soul. The temporal creation was however, a different creation. The first man (the lower Adam) was a physical or temporal entity that was **made** by God, the son, and other spiritual beings, after having evolved to a certain point. He was **made** into a new creature, having already existed. He was perfected physically to look like his creators and given a soul. The word **create** means to bring into being or existence and the word **made** (a past tense of the word **make**) means to construct. Man on this planet, the temporal creation, was remade or constructed, having already existed. He was patterned after the *upper Adam* when he was given a soul (his spiritual nature) and made to look physically like his creators.

393

To summarize briefly: **To create** is to bring something into existence. **To make** is to construct something from material already existing. The upper Adam (the spiritual Adam) was **created.** The lower Adam (the temporal Adam) was **made.** Isn't it amazing how definitions of the simple words, **create, make,** and **made,** change the whole meaning of texts? So similar, yet profoundly different when analyzed.

Religious leaders have reasoned that the two contradictory creation accounts were the result of two different authors who wrote as they perceived the creation and placed each of their similar but different accounts in *Genesis.* But, once the concept of two different creations is understood then the two accounts of the creation in *Genesis* make sense.

Now, the timing is the subject of much disagreement. But, there are two main themes that permeate every known religion that involves the origin of man. One theme is that a race of man was created on the temporal earth and was allowed to proliferate much like any other creation, except with superior intellect. Tools, pottery, clothing, tombs, and communities of evidence can substantiate his existence many thousands of years prior to the great flood mentioned in the history of every ancient religion. He migrated. He hunted and in many cases farmed and domesticated animals.

The other theme present in every cultural religion is that there was a spiritual creation of man. This man was highly intelligent, strong, and genetically pure. He also had a soul and an inextricable link to one Creator God from the very beginning. A grand separation occurred between those that held to their divine nature and those who did not. The faithful and pure survived and at some point in time—one of the main purposes of your authors—we intend to contribute the idea that the temporal and spiritual races came together here on this planet. The results changed everything in the universe.

The *Sefer Yetzirah* or the *Book of Creation* states that "God allowed the universe to develop by itself (the temporal universe after its initial creation). All the laws of nature and the properties of matter had been fixed for all time, as it is written, *"He has established them forever. He has made a decree which shall not be transgressed"*[6] It is similarly written, *"Whatever God decrees shall be forever; nothing shall be added to it, and nothing shall be taken away. "*[7]

"....... around some 25,000 years ago, man developed all the physical and mental capabilities that we possess today. This man had evolved from *"the dust of the earth"* (the temporal earth)[8], but he still lacked the divine soul that would make him a spiritual being. God then created Adam (the lower Adam), the first true human being with a soul (on the temporal earth), *"and He blew in his nostrils a soul of life"*[9]

This ancient book with rabbinical interpretation, states that man evolved from *"the dust of the earth."* This could mean that he evolved, just as scientist say, from an early branch of the primate family, who evolved also from *"the dust of the earth."* The Zohar speaks of early man having ape like characteristics. Apparently certain laws were put forth on this temporal planet to allow things to come forth or evolve by themselves without further interference except when necessary. When the man-like creatures had evolved to a certain point then God, the Son, along with other spiritual beings, made a new being with a soul, which was in their image. It is the same process that God, the son, has done in countless other worlds in this temporal universe, that he created.

There is an ancient Babylonian text that gives a good description of how the temporal man may have been made, called the *Atrahasis*. It was written no later than 1700 B.C. In essence the god Anu and Enlil, his son, called a great meeting of all the gods and goddesses. Enlil addressed the meeting to try and avert a rebellion that one is led to assume that it may have been Enki, Enhil's brother. Now Anu, having defeated the god Alalu, was the supreme ruler of some of the gods and the Annunnaki (fallen angels or watchers).

The Annunnaki were also present at the meeting and complained of the heavy workload that they had to bear. Enki spoke up, saying that he had a solution:

"While the birth-goddess is present,

Let her create *Lullu,* man

Let him bear the yoke....

Let man carry the toil of the gods."

Enki's suggestion was unanimously accepted and the gods voted to create the worker. "Man shall be his name," they said. Enlil called upon Mami (Ninhursag), his sister to help in man's creation. Enlil said to her, "You are the birth-goddess----create workers!" She said that it is not possible for her to make things but that she would need the skilled help of Enki. The story depicts Enki as a chief scientist among the gods who is learned in the science of genetics.

The text continues with the slaying of the god, "We-ilu, a god who has intelligence," from whom man was to gain intelligence. Enki and Mami, the mother goddess, created man from clay and from the flesh and blood of the slain god.

"From his flesh and blood

Let Nintu (Mami), mix clay

That god and man

May be thoroughly mixed in the clay....

Let there be a spirit from the god's flesh."

A surrogate mother was needed for the birth process. Enki offered the childbearing service of his own wife, the goddess Ninki.

"Ninki, my goddess-spouse, will be the one for labor.

Seven goddesses-of-birth will be near, to assist."

In a hospital-like place, called the house of destiny, the gods were waiting for the procedure and birth to take place. Finally, success was achieved and the birth of the first *Lullu* or man took place.

There is a well-known cylinder seal from ancient Mesopotamia showing a laboratory with flasks, a burner, and the god Enki holding a test tube. Mami is shown mixing a concoction on a burner and the goddess Ninki, is shown presenting Adam to Enki.

The Aztec creation of mankind:

Quetzalcoatl descended to the underworld to obtain from its ruler, Mictlantecuhtli, bones and ashes of the previous generation of humanity, from which to re-create (remake) man. Quetzalcoatl delivers his precious cargo to the gods assembled in Tamoanchan. Tomoanchan is the place of creation and origin of man. It is the Aztec Garden of Eden.

Eihuacoatl and Quetzalcoatl grind the bone fragments into a mass and place them into a precious vessel. The gods then perform auto-sacrifice, permitting their blood to drip thickly on the mass of ground-up bones. After 4 days a male child emerges, and after 4 more days, a female infant emerges.

There are several other Mesopotamian and many similar creation myths world wide that speak of the gods mixing their blood with clay or pulverized bones from the preceding generation, even using the flesh of a sacrificed god to create man. Perhaps this was one way to convey to man that he was a special divine creation.

As we, the authors view the above early texts, it would seem that early man evolved to a certain point, and then the Gods came down and genetically re-engineered him into their image. Whether this process took days, years, or thousands of years to accomplish matters little as time is of no particular essence to the gods on the temporal creation. They measure time in the greater eternal prospective. Essentially this is what is conveyed in *Genesis* 1:26 in the temporal creation account:

"And God said, Let **us make** man in our image, after our likeness...."

Man had already evolved to a certain point but lacked the image of the Gods. The *Sefer Yetzirah* refers to them as soul-less beings. Early hominids may have evolved directly from the dust and not as an early off shoot and descendant of the primates because "kinds" of one animal do not evolve into another "kind", however, they could possibly be genetically re-engineered. So, early hominids could have evolved directly from the dust or descended from an early off shoot of the primates, who may have also evolved from dust. Either way, the early hominid was not fully human until the Gods intervened.

The following statements are taken from the website www.parentcompany.com/creation_essays/essay42.htm:

"The currently hot evidence for human evolution comes from molecular biology. The DNA molecules or genes of different species are compared, and particular protein molecules are likewise compared. The more similar are the sequences of nucleotide units in the DNA molecules, and the more similar are the sequences of amino acid units in the protein molecules, the closer the evolutionary relationship is assumed to be.

There are problems with this theory of DNA and protein evolution, however. Different proteins sometimes give different results from each other and from DNA. For example, the cytochrome *c* molecules of man is identical to that of chimpanzee, but differs by one amino acid from that of the rhesus monkey considerable more distant from man, but chimp still quite close. In fact, it has been said that chimp DNA is as much as 99 percent similar to human DNA.

But, then it is reasonable to ask why there is so much difference between chimps and humans. It appears that DNA may not determine everything after all. One thing certain is that DNA comparisons do not prove evolution."

What appears to be clearly confirmed is that many forms of humanoid creatures lived thousands and even millions of years ago, both giants and dwarfs. The bones of *Gigantopithecus*, discovered in 1946, date back to nearly 10 million years. This giant stood more than 8 feet tall and weighed 400 to 500 pounds. *Ramapithecus* an ancient primate, was small, only a few feet high. It is now thought that both these possible "missing links" are an evolutionary side branch and were contemporary with our true ancestor.

Current scientific opinion is that a primate named *Australopithecus* or "Lucy" is the best candidate for a direct ancestor of man. This species dates back nearly 4 million years. About 115 thousand years ago modern *Homo sapiens* appear abruptly on the fossil record. While clearly a derivative of archaic *Homo sapiens*, the change is sudden. Archaic *Homo sapiens* is a derivative of *Homo erectus*. Modern *Homo sapiens* appeared on the scene **before** the more primitive Neanderthal man who are the further natural evolution of *Homo erectus*.

The fossil record supports the belief that modern *Homo sapiens* was indeed a special creation and did not evolve naturally from primitive forms because his appearance on the scene was sudden.

In summary, the ancient texts and fossil records indicate that early hominids, in the temporal creation, either evolved from the dust or descended from a very early primate branch, who also evolved from the dust. It appears that the gods came about 115 thousand years ago and genetically improved or re-engineered *Archiac Homo sapiens* into *Modern Homo sapiens* who were made into a new creation, patterned after their image; patterned after the gods. Therefore the evolutionists are correct in stating that man, on the temporal creation, evolved and the creationists are also correct in their belief that man was made. Modern man was an evolved genetically re-engineered special new creation on the temporal earth. **He was both!** (Don't you feel like breaking into a stanza of the *Hallelujah Chorus* here? Finally we can all get along.) **And** the most amazing realization is that we have had the answer to the creationism versus evolution all along in the biblical scriptures...... the meaning and use of the words **create, make,** and **made** being hidden in plain view in the *Genesis* Creation texts.

Now as to the matter of the races, there are legends that still prevail in Central America, South America, and in the southwestern Native American tribes, that the gods created four different races in their own image, in various parts of the world, at the same time. Their belief was the Red race was created in the Americas, the Yellow race in Asia, the Blue (Black) race in Africa, and the White race in Europe. Each race represents one of the four divine cosmic forces as elements; the Red race, earth elements; Blue race, fire elements; White race, air elements; and the Yellow race, water elements. *The*

Zohar states that, "race is the product of the environment," which is dependent on location, sunlight, food, air, and availability of outside marriages for genetic diversity. Genetics definitely play a role in the different ways people appear. For example, Native Americans have a gene that puts their capillary blood vessels close to the skin surface that gives them a red cast, or coloration. Orientals have a gene that puts a layer of fat just under the skin that produces the yellow coloration. The amount of skin pigment or melanin determines how dark or light a person's skin complexion will be. It is really all in the inherited genes and there is a great variety of genetic diversity among men and animals. Scientist today feel that there are no distinct races of people because there is diversity of color and traits even in a "race." Race was known only on the temporal creation. We will probably never know for sure on the issue of race. Environment definitely is a factor, producing genetic diversity, but whether the gods created the races is speculative at most.

Meanwhile, back to Mu. Just suppose that Mu, as the South Pacific legends say, is the motherland or cradle of civilization. Could not early hominids have migrated out of Africa by crossing land bridges that may have connected Africa to Australia in remote times when the sea levels were much lower?

Once arriving in Australia, they found the continent and close fertile island lands, off shore of Australia, a desirable place to live; a sort of Garden of Eden on this planet. Then about 115 thousand years ago the gods—God, the son, and other spiritual beings—came to the island paradise and re-engineered them into a new creation that looked like them in every aspect. Is it not unreasonable to think that cataclysms made it necessary for one or more major migrations from Mu to cross existing land bridges back into Africa, Asia, and Europe? Isn't it much more likely that the reason all these cultures have common symbols and workings in their religions and architectures is that they have a common origin at Mu?

This commonality is irrefutable. Each year the evidence grows that anciently there was a very modern and mobile civilization here on this planet. Their knowledge was astounding. Their goodness and benevolence was greater than anything modern man has achieved. Unfortunately, their evil was also as proportionately astounding.

The advent of this knowledge was sudden and profound. There is only one explanation for which any evidence remains. This knowledge was brought in from somewhere else by an extremely advanced and purposed source. Once modern man began to multiply, the rebel fallen angels came to corrupt the seed of men on this planet, on orders from Lucifer. They found the daughters of these men to be fair, seduced them and they bore giant offspring. The rebel watchers—fallen angels—or Nephilim, also called *The Elder Gods*, were super intelligent beings; fallen sons of God. They co-existed with men, teaching man all manners of science, farming, art, music, writing, maritime skills in boat building and navigation, medicine, metal works, and architecture.

Man viewed them as their gods, worshipped, and paid them tribute. Ancient writings, hieroglyphics, and archeological evidence support this very well. Even Buddha claims in his writings that he associated with the *Elder Gods*. Man was a willing worker for his gods. These off-worlders were capable of such wickedness. Many of the watchers deliberately taught them sexual immorality, vanity, murder, mayhem, war, and idolatry. They were also, perhaps in an effort to redeem themselves with lifetimes of good works after their abominations, capable of incredible good and glorious technology, which some freely shared with their subjects. Together, the watchers and man developed a highly civilized and intellectual civilization that eventually spread colonies into South America and colonized other parts of the world across the oceans.

Although Mu has been portrayed as a peaceful society governed by laws and regulations, sexual perversions and immorality were standard everyday fare that was not only permitted but expected as normal. For example, the giant offspring practiced sodomy. To be chosen to be sodomized by one of them was considered an honor, to refuse was to suffer the retribution of dishonor. This is an obvious and purposeful deviation from the procreative functions of the human body. It was used by the followers of the order of Cain to consummate a covenant to commit murder for personal gain and an initiatory rite for entrance into secret combinations of evil practiced by the Nicolaitan's and the Gadianton Robbers. The name is derived from the city of Sodom that was destroyed by God in the *Old Testament* for completely corrupting the holy act of copulation.[10] The practice has been determined to be one of the most medically unsafe acts that one human can perform with another. The transfer of disease is so efficient that of those people that regularly practice sodomy, only two percent on average live long enough to collect social security.

According to Churchward, South America looked different in those days prior to the cataclysm. The Andes were mere hills and there was a large inland lake or sea that filled most of the area of northern South America with an opening channel or river to the Atlantic. Tiahuanaco was first built on the western rim of this inland sea and was a major sea port. By 10,000 B.C., well established colonies were in parts of South America. The Nephilim, or the *Elder Gods,* and their giant offspring were said to be the builders of Tiahuanaco in Bolivia, when the site of the original city was at sea level. In addition, legends say that they were the builders of the 26 stone cities of Akakor. (readers are referred back to the chapter, *The Giants*, for more details)

It is further said that *The Empire of the Sun* evidently, had an emperor who acted as a figurehead for their solar deity whose symbol was "The Sun." This emperor bore the title of "Ra Mu" and he presided over a group of priests called the Naacals.

Lemuria sent high priests called the Naacals or holy brotherhood to various colonies to teach their doctrines and preserve writings of the history of this planet and knowledge. Science and religion were covered as one subject. One colony of Naacals officiated in the great *Temple of the Sun* in Tiahuanaco. It was a Solar Brotherhood that became better known as the "Great White Brotherhood" on earth. It was called a white brotherhood, not because of race, but because they believed themselves to be spiritually enlightened; the color white being associated with spirituality. At one time a great golden sun disc, encrusted with precious gems, hung in the *Temple of the Sun.* The disc had the image of a man's face in the center with seven light rays radiating from around the golden orb that represented the sun. Several copies of the golden sun disc were placed in other temples, the original being in the *Temple of the Sun* in Mu.

According to legend, members of the Naacal brotherhood left Mu for Egypt, Burma, India, Tibet and Atlantis, about 30 thousand years ago. In the Hindu epic of *Ramayana*, Valmiki mentions the Naacals as having arrived in Burma from the lands to the east. Rutas appears to be the Hindu name for the Pacific continent of Mu. Egyptian records state that the first Egyptians came from the Southern Sea in large sailing ships. The Southern Sea would be the Red Sea today, but 30,000 years ago it probably was the Indian Ocean. (Readers are referred to *The Brave New World* chapter for explanation) It empties into the Indian Ocean that washes up on the western shores of Australia, the last remaining remnant of the third island continent of Mu today. In those days Egypt was a fertile green Nile river valley, having much rain. Where-ever the Naacals went, they built pyramids and temples, and established knowledge among modern men that had previously dispersed earlier to many parts of the world. Historians will someday admit that civilization is thousands of years older than previously thought. Then they will have to acknowledge that men have been sailing and mapping the world's oceans for millennia. With travel comes colonization and established trade routes.

It would seem that Lemuria was an idyllic civilization but it wasn't, it was flawed in the creator's eyes. The seed of man had been corrupted to the point that man had forgotten who his true creator had been and acknowledged other gods. In God's eyes, this is the ultimate insult! The cataclysm of 9,500 B.C., was sent by God to destroy the corrupted temporal earth. The Fallen Angels could see into the future with their "seer stones" and fled earth before the cataclysm but most of their giant offspring were destroyed.

Churchward states in his book, *The Lost Continent of Mu*, the following:

"The record of the destruction of Mu, the Motherland of Man, is a strange one, indeed. From it we learn how the mystery of the white races in the South Sea Islands may be solved and how a great civilization flourished in mid-Pacific and then was completely obliterated in almost a single night."

The Nephilim were responsible for the horrible destruction of the Motherland known as Mu. All that remains today are the volcanic mountaintops in the South Pacific and possibly Australia.

The next chapter, *The Lost Civilization of Atlantis*, will describe some of the other colonies of Mu that existed before the cataclysm and learn how these lost civilizations continue to affect us, even today.

End Notes

[1] *(The Lost Continent of Mu, Colonel James Churchward)*

[2] (Easter Island Tablet, Greek Record, Troano Manuscript, S.A. Record, Indian and Mayan Records, quoted by Colonel James Churchward in *The Lost Continent of Mu*)

[3] (Churchward quoted the Codex Cortesianus, an ancient Mayan Manuscript, Troano Manuscript, and Lhasa Record of Tibet)

[4] Excerpts taken from *The Return of The Serpents of Wisdom* ©1997 by Mark Amaru Pinkham. Published by Adventures Unlimited Press, Kempton, ILL. All rights reserved. Used by permission of the publisher, p. 48)

[5] (*The Origin and Destiny of Man*, Lytle Robinson copyright 1972. pg. 122)

[6] (*Psalms* 148:6)

[7] (*Ecclesiastes* 3:14)

[8] (*Genesis* 2:7)

[9] (*Genesis* 2:7). (Excerpted from *Sefer Yetzirah* by Aryeh Kaplan ©1997 with permission of Red Wheel/Weiser, York Beach, ME and Boston, MA pp.186-187) Parenthesis added.

[10] (PoGP *Moses* 5)

The Lost Civilization of Atlantis

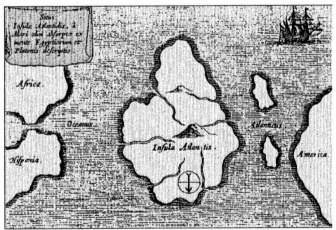

Map by Athansius Kicher 1602-1680. Latin says, "Site of Atlantis now beneath the sea according to the beliefs of the Egyptians and of the description of Plato." Note the arrow pointing to what we now denote as South.

Atlantis was first colonized, maybe as early as 70,000 years B.C., with the Naacals and modern man who were from the Motherland Mu, having arrived in ships. The first historic mention of Atlantis was made by Plato, scholar, historian, and philosopher, who lived 400 years before Christ. Plato wrote an account in the *Timaeus* about his uncle, Critias, who told how his ancestor Solon (639-559

B.C.), a great statesman, had visited Egypt nearly two centuries ago. Solon, being a historian, desired to know more about the ancient history of mankind and of Egypt, even before its establishment as a kingdom, as he realized that the Egyptians had a reputation of knowing far more ancient history than the Greeks. He wanted to preserve these records of antiquity for his fellow countrymen.

In *The Atlantis Blueprint,* Wilson and Flem-ath state the following: Solon was told by an old Egyptian priest, "...and out in the ocean, beyond the Pillars of Hercules (which we now know as the Straits of Gibraltar), there was an island-continent called Atlantis, as large as Libya and Asia combined. (By Libya he meant all of North Africa and by Asia he meant an area equivalent to the Middle East.)

The priest then made a baffling statement. From Atlantis, he explained, it was possible to reach other islands that formed part of Atlantis, and from them, the whole opposite continent that surrounds what can be truly called the ocean." Wilson and Flem-ath further state in their book, "the Greek Neoplatonist philosopher Proclus had stated that Plato's student Crantor (c. 340-275 B.C.) had visited Egypt, where he saw pillars inscribed with the legend of Atlantis."[1]

Plato wrote in another account called the *Critias,* of how Atlantis was founded by the sea god Poseidon (Neptune), who married a mortal woman and fathered five sets of twins. Poseidon gave the island its name of *Atlantis,* and also named the surrounding sea the *Atlantic Ocean.* His account states, "The god built her a home on a hill, and surrounded it with concentric rings of sea and land. The twins were each allotted a portion of the island, and over the generations extended their conquests to other islands and to the mainland of Europe.

Great engineers, the Atlanteans built a circular city, 11 miles in diameter, with a metal wall and a huge canal connecting it to the sea. Behind the city there was a plain 229 by 343 miles wide, on which farmers grew the city's food supply. Behind this were mountains with fertile meadows and every kind of livestock, including elephants."

The *Critias* has many pages describing the life style of the Atlanteans and their social structure. It speaks of magnificent buildings, temples, and palaces constructed of many-coloured stones, of hot and cold fountains, indoor plumbing, communal baths and dining halls. They built numerous canals and covered bridges. Temples were covered with gold and silver, both inside and outside. Roofs inside of temples were ivory and the statues were of gold. One temple contained a colossal figure of Poseidon standing in a chariot, drawn by six winged horses, of such size that his head touched the roof of the building. On the outside of the temple were ten statues of gold that represented the ten kings and their wives.

Plato describes the harbors of Atlantis as being filled with ships and vessels from all parts of the world, engaging in trade and commerce. The royal docks were lined with shops and naval stores. They wanted for nothing. The Atlanteans, and especially the Poseidians, studied the creative energies of the universe and penetrated the essence of nature's storehouse; the vibrations of plants, jewels, and metals—the latter including their vibratory effect on the psychic and intuitive nature of man. Their technology was highly advanced, calculating the meaning of numbers, the stars, and the elements.[2]

From the narrative we learn that the island was divided into ten parts, each part ruled by a king, and each part had a capital city; a sort of United States of Atlantis. All matters pertaining to government were deliberated until agreement was reached.

Solon was told by the old priest that his knowledge came from the sacred Egyptian registers that were then 8,000 years old, or approximately 8,600 years B.C., of Plato's time. A written account in verse was made by Solon and handed down to Plato who preserved it in the form of a dialogue for posterity. Unfortunately, he started the dialogue late in life and left the story unfinished.

Plato further states, in his narrative, that the old priest of Sais told Solon the following, "Now in the island of Atlantis there was a great and wonderful empire which had rule over the whole island and several others, as well as over parts of the continent; and besides these, they subjected parts of Libya within the Columns of Hercules as far as Egypt, and of Europe as far as Tyrrhenia."

According to Roman and Greek mythology, Poseidon (his Greek name) or Neptune, as he was known by the Romans, was one of the Olympian gods. He was the son of the Titan god, Kronus and the Titanide, Rhea. The Olympian god Zeus (Jupiter to the Romans) was his brother.

The Titans, also known as *The Elder Gods*, were an ancient race of gods in Greek mythology, which ruled the earth before the younger Olympian gods overthrew them. In Greek and Roman mythology they reigned for untold ages as supreme rulers in the universe and were of enormous size, possessed super intelligence, and were endowed with incredible strength. When they came to Earth, this planet was divided up between them, each Titan ruling a certain portion. The ruler of the Titans was Kronus (Saturn to the Romans) who was defeated and dethroned by his son Zeus. The Titans were the children of Uranus (Heaven) and Gaea (Earth). Uranus feared his giant children, and chained them down in Tartarus, which was far under the earth.

After Zeus (Jupiter), having defeated his father Kronus (Saturn), he ascended to the throne. Kronus fled to Italy and brought in the Golden Age, a time of peace, happiness and plenty, which lasted as long as he reigned.

Does this not sound familiar? Could the Greek and Roman gods of mythology have been actual historical figures? Remember legends state that the Elder Gods had six fingers and six toes and were the builders of Tiahuanaco in Bolivia. These were the Nephilim. The Titans were a race of Elder Gods, which race has been previously identified as Nephilim, thereby the Titans were the fallen sons of heaven (Uranus or God) and the Olympians were their giant offspring. The Greeks also referred to these Elder Gods as Els, and Elanakim, short for Elder Race or simply the Giants. Many statues of these gods and goddesses were depicted as having wings of angels, some colossal in size. The *Book of Giants* of the *Dead Sea Scrolls* states that the Nephilim were giants and even had wings. These facts alone indicate to us that the Greeks knew these gods were Nephilim and that indeed their incredible myths and fables were based on real historical figures and actual events. Men accepted these beings unconditionally as their gods because they had lost all knowledge of their true maker. As far as they were concerned, their gods were their creators. Note, in the Depiction of an Elder God below, the six fingers on each hand which was a defining trait of the Elder Gods or Nephilim.

Depiction of an Elder God

Titanides! Here is the first mention of female fallen angels, who were called goddesses. *D&C* section 29:36 states,

"And it came to pass that Adam, being tempted of the devil—for, behold the devil was before Adam, for he rebelled against me, saying, Give me thine honor, which is my power; and also a third part of the hosts of heaven turned he away from me because of their agency."

Many sacred scriptures and ancient writings say that there was a Great War in heaven. We interpret this to mean war in the highest spiritual universe and that one third of the host of this spiritual universe fell or were driven out to the temporal universe. Whether or not the other spiritual universes were involved in this conflict is not clear but the gnostic writings indicate that they were affected and may have participated. One third is a huge number. It could number in the tens of billions or more! Because of their vast numbers is it not logical to conclude that not all of them came to planet Earth? Common sense would also dictate that they were not all-male entities.

Chief of the tribe, Tatunca Nara described that the *Chronicle of Akakor* is "a history of the oldest people in the world..." He explains that for a long time the tribe existed in a highly primitive state until the gods came and brought them light. On earth these beings ruled from Akakor, erected around 14,000 years ago. The chief went on to say that their home planet, called Shwerta, lay in the depths of the universe: "a mighty empire consisting of many planets as numerous as particles of dust in the street."

As previously discussed the builders of Akakor and Tiahuanaco were Elder Gods or the Nephilim. Could planet Earth have been a mere outpost? Maybe in the beginning. It appears that they were looking for the planet that the Redeemer might or could one day be born into.

Assyrian winged Nephilim

This wasn't made known until several thousand years later when the spirit earth united with this particular planet and Noah, a descendant of Adam, made his appearance. Until then perhaps they were colonizing all possible planets onto which the descendant of Adam (Noah) could come. In theory corrupting the seed of all these planets would stop the birth of the Redeemer, because they knew his mission, if accomplished, would redeem **all** fallen planets (worlds) in their possession. *John* 3:16 somehow became twisted in translation because in the original Greek, the word for *world* is the Greek word *cosmos.* The Greek word *cosmos* refers to anything and everything that is in this universe, therefore the entire universe would be redeemed. If this earth were the only fallen planet it wouldn't have been necessary to redeem the entire universe. They erred by looking for an existing race not knowing that the seed of Adam would be literally transported to some planet in this universe. (Do we see a sequel to this book here?)

By some accounts, the Titan god Prometheus and his Titan brother Epimetheus, were delegated by Zeus to create man, at least that is what they probably told man when he inquired as to who was his creator. Most of the Greek and Roman stories indicate that men were treated well by these gods but, at the same time, they were at their mercy. They usurped the original commandment given to man *subdue the earth* with one of their own design; *subdue one another.* What better way to hold man in willful submission than to tell him that they were his creators and gods. No self-respecting Nephilim in his right mind would have told man that they had been kicked out of Heaven by God, and that his leader was Lucifer whose mission was to corrupt their seed and take over planet Earth. It does appear that the Nephilim on the temporal creation were much less vicious than those from Enoch's account on the spiritual creation. Why? Maybe, it is because Adam and his descendants down to Noah on the spiritual creation were very intelligent beings, having much knowledge of the workings of God and of the creation, were more difficult to corrupt. On the other hand, men on the temporal creation, although intelligent, were uneducated, simpler, possessed trusting childlike qualities that made them easy marks for corruption by these impressive supernatural beings.

Plato's description of the Atlanteans is one of a nation united in peaceful endeavors, caring for one another, prizing intelligence and knowledge above material assets. Atlantis in some ways was an utopia but one only has to read some of the stories preserved in myth about the activities of the Titans and Olympians, which has filled books, to realize that they were also war like and notoriously promiscuous with mortal women and animals. Some of their unions with animals produced monsters. The centaurs were half human and half horse. The fabled Minotaur was half human and half bull. The demigod Pan had the legs, ears and horns of a goat. Satyr's were demigods with the tail and ears of a horse and body of a man. Gorgons were snakey haired women (Medusa) whose frightful appearance turned the beholder to stone. Sphinx's had the head of a man and the body of a lion. There were bird men, men with the head of various kinds of birds, namely falcons and ibis. The Chimaera was a monster composed of the head of a lion, the body of a goat and a serpent for a tail. Griffins had the body, tail, and hind legs of a lion, and the head, forelegs, and wings of an eagle. Some of these creatures had the status of gods and demigods being the offspring of the Nephilim.

The Book of Jasher, p. 101& 102, gives an account of Anah, son of Zibeon who was the son of Seir the Horite, who found the Yemim in the wilderness when he fed the asses of Zibeon his father. The land of Seir is the dukedom where Esau lived and where his children were born. The account of the Yemim reads as follows, " And whilst he was feeding his father's asses he led them to the wilderness at different times to feed them. And there was a day that he brought them to one of the deserts on the sea shore, opposite the wilderness of the people, and whilst he was feeding them, behold a very heavy storm came from the other side of the sea and rested upon the asses that were feeding there, and they all stood still.

And afterward about one hundred and twenty great and terrible animals came out from the wilderness at the other side of the sea, and they all came to the place where the asses were, and they placed themselves there. And those animals, from their middle downward, were in the shape of the children of men, and from their middle upward, some had the likeness of bears, and some the likeness of the keephas, with tails behind them from between their shoulders, reaching down to the earth, like the tails of the ducheephath, and these animals came and mounted and rode upon these asses, and led them away, and they went away unto this day. And one of these animals approached Anah and smote him with his tail, and then fled from that place. And when he saw this work he was exceedingly afraid for his life, and he fled and escaped to the city. And he related to his sons and brothers all that had happened to him, and many men went to seek the asses but could not find them, and Anah and his brothers went no more to that place from that day following, for they were greatly afraid for their lives."

The practice of bestiality with mortal men will produce nothing as offspring, but this same practice with spiritual beings and mortal animals did produce monsters. The mating of the Nephilim with mortal women produced giant offspring. Subsequent generations, over many thousands of years, became as the *Critias* states, "diluted with too much of the mortal admixture and the god-like element or divine portion of the Atlanteans began to fade away." Human nature many times took control and got the upper hand making these generations appear base and degenerate. For example, about 35,000 years ago Cro-Magnon man suddenly "invades" the western shores of Europe and North Africa. [3] Professor J.L. Myers (1923-1939) describes Cro-Magnon culture as "a well-marked regional culture of the Atlantic coastal plain." The terms "Atlantic" and "Paleo-Atlantic" are often used when referring to Cro-Magnon culture. Colonies of Cro-Magnons are always clustered in the west with the number of colony sites diminishing towards the east. To the west, there is nothing today but empty ocean. Where from the west did they come? It is a mystery that continues to plague anthropologists.

The clue to their origin is their size. Cro-Magnon man was much larger than today's modern man but was fully modern in appearance. His average height was 6 feet 6 inches with many much taller. He was bigger and stronger than today's men. Even his brain size, which averaged from 1617cc to 1743 cc, exceeds that of men today. He was very intelligent. We would classify them in our times as giants. Giants that probably were degenerate descendants of the giant offspring of the Nephilim; degenerate in the respect that subsequent generations became smaller. When the DNA of recovered biology is tested, it becomes evident that modern man could not have descended from Cro-Magnon man. It is genetically impossible. They were mortal but their size and strength was a throw back to their now faded divine portion. In their day Atlantis was located westward in the Atlantic Ocean.

Atlantis had many colonies of peoples in the world that was regularly serviced by sea faring routes. These sea faring routes were established by the use of maps or charters. An ancient map has been discovered that shows the Antarctic without ice during the Golden Age. Hapgood, author of *Maps of the Ancient Sea Kings,* discovered a map, in the Library of Congress, made by Oronteus Finnaeus in 1531, showing Antarctica much as it does today without the ice. Obviously, this map was not the original but could be a copy passed down from the Golden Age times. Hapgood uncovered another map showing Siberia and Alaska joined with no Bering Strait. A land bridge once joined the two countries more than 12,000 years ago. Historians have not accepted the idea that the world had been mapped prior to 10,000 B.C. So, Hapgood has been largely ignored.

The giant descendants of the Olympians simply were some of the Atlantean colonizers of Western Europe and the African coastal plains. They brought their advanced tool industries and technology with them from Atlantis. There is no evidence of a "gestation period" for tool development in their colony sites. Advanced tools and their extensive use appeared suddenly in the archeology, without the presence of any precursors whatsoever. In other words, there were no prototypes or simpler tools that developed into more advanced tools. About 16,000 thousand years ago another similar race colonized the continent and about 10,000 years ago a third race known as the Azilians invaded the same continent, all originally from Atlantis.

Atlantis had several destructive periods. A portion of it sank about 26,000 years B.C., the same time that one of the three island groups sank in Mu. This was probably due to worldwide tectonic plate activity that increased volcanic eruptions and earthquake disturbances of cataclysmic portions. With a decrease in land size, it became necessary to colonize other areas for population control. Every few thousand years Atlantis underwent serious upheavals and became geologically unstable forcing hordes of refugees onto the Atlantic islands and western coasts of the continents. Each invasion was more advanced than the last, happening over a period of just under 25,000 years, namely, the Aurignacian, Solutrean, Magdalenian and Azilian cultures.[4] The Azilian culture appeared just before the final destruction of Atlantis and since the submerging of that land there have been no more Cro-Magnon invasions because Atlantis was the apparent source of the invasions, now long gone.

An anthropological fact usually ignored in some factions is that there are notable differences in types of paleolithic man. The terms Cro-Magnon and Modern Man are used as if they were synonymous, but in a strict sense they are not. All Cro-Magnons are Modern, but all Moderns are not Cro-Magnon. The Cro-Magnon culture disappeared after the 9,500 B.C., cataclysm. The Moderns survived thus the sudden decrease of the cranial capacity (average 1450 ccm) seen after the cataclysm.

If nothing else, the authors wish to dispel the notion that the Cro-Magnon cultures were a "primitive and boatless" society. They were intelligent descendants of a highly civilized society who became pioneer colonizers on the Western European and African continents, having arrived in ships. A number of Cro-Magnon village sites show evidence of agriculture dating back as far as 15,000 B.C. They lived in constructed houses, had all manner of advanced tools, needles, obsidian knife razors, musical instruments, even bone calendars whose symbolic notations border on writing.[5] His caricature artwork has shown man wearing shoes, woven fabric pants, coats, hats, and clean-shaven [6] They were not shaggy-haired savages wearing animal skins. They even domesticated several species of animals, which may have included the horse, although the god Poseidon is credited with taming of the horse.[7] With each period of colonizing invasions came new innovations in technology carried from Atlantis.

According to Churchward, the Nagas (Naacals) left Mu, came to Burma and then entered, colonized, and founded *The Rama Empire* or what is now known as India, Burma, and Iran. Rama was the emperor of *The Rama Empire.* He acted as a figurehead for their deity whose symbol, like that of Mu, was *"The Sun."* The capitol city of Rama was anciently known as Deccan, but today called Nagpur. In the Rama Empire were seven capitols, which became known as the "Rishi Cities." The name Rishi in Sanskrit means "Master" or "Great Teacher."

Atlantis and the Rama Empire had much in common as both were colonies of Mu, both developed advanced civilizations with high technology, higher than the civilization of Mu, and both had airships.

The Atlanteans called their airships Valixi and in the Indian Epics the airships of Rama are identified as Vimanas.

A photo of Vedic glyphs. Note the helicopter and spacecraft.

The Atlanteans applied their technology toward military applications while the Rama Empire always used their advancements towards peaceful endeavors. They also used flying craft in their engineering to lift large structures and stone into place. Vedic texts indicate there existed three types of aircraft; one for personal transportation, one for lifting and transporting heavy loads, and one for battle.

Besides having airships the Atlanteans developed weapons of mass destruction. They had mastered the use of crystals and the power of the sun. In addition to being able to supply power, light, and communication, they were capable of extremely powerful destruction. Subtly used, they could be adapted as a means of coercion, torture and punishment. One weapon was described as a fireball or "Kapilla's Glance" that could reduce a whole city and its inhabitants to ashes in seconds. Another weapon is described as flying spears (missiles) that could ruin entire "cities full of forts." The agenda of the Atlanteans was to rule the world by subjugating its colonies and other colonial empires of Mu. At the height of both civilizations, war broke out between them.

According to the *Lemurian Fellowship* lessons the Atlanteans invaded the Rama Empire with a well-outfitted army, equipped with various types of formidable weapons. One Atlantean leader landed their Vailxi, backed up by the invading army, outside one of the Rishi cities and sent a message to the Rama Priest-King to surrender or be destroyed.

The Priest-King answered, "We of India have no quarrel with you of Atlantis. We ask only that we be permitted to follow our own way of life."

The Atlanteans perceived their reply to be one of weakness as they knew that the Rama Empire lacked weapons of mass destruction. The Atlantean general sent another message back to the Rama Priest-King, "We shall not destroy your land with the mighty weapons at our command provided you pay sufficient tribute and accept the rulership of Atlantis."

Trying to prevent war, the Priest-King humbly replied, "We of India do not believe in war and strife, peace being our ideal. Neither would we destroy you or your soldiers who but follow orders. However, if you persist in your determination to attack us without cause and merely for the purpose of conquest, you will leave us no recourse but to destroy you and all of your leaders. Depart, and leave us in peace."

Confident that the Indians lacked the military power to stop them, the Atlantean army began marching at dawn on the Rashi city. As the army approached, the Priest-King raised his arms toward heaven and applied a sort of mental telepathy, known only to certain yogis in the Himalayas today, aimed toward the general and officers of the Atlantean army, in descending order of rank, that caused them to instantly fall dead in their tracks. Seeing the power of the Priest-King, the army panicked, turned tail and fled to their waiting airships and retreated back to Atlantis. Humiliated by their defeat the Atlanteans decided to annihilate the major Rashi cities and take the empire by force by using weapons of mass destruction. It wasn't until the atomic bombs were dropped on Japan that Sanskrit scholars could fully understand what was described in the Indians Epics.[8]

The following are actual verses taken from the Indian Epics:

"Gurkha, flying a swift and powerful vimana,

hurled a single projectile

charged with all the power of the Universe.

An incandescent column of smoke and flame,

as bright as ten thousand suns,

rose with all its splendor.

It was an unknown weapon,

an iron thunderbolt,

a gigantic messenger of death,

which reduced to ashes

the entire race of the Vrishnis and the Andhakas.

The corpses were so burned

as to be unrecognizable.

Hair and nails fell out;

Pottery broke without apparent cause,

and the birds turned white.

…After a few hours

all foodstuffs were infected…

…to escape from this fire

the soldiers threw themselves in streams

to wash themselves and their equipment."

The Mahabharata

"(It was a weapon) so powerful

that it could destroy the earth in an instant—

A great soaring sound in smoke and flames---

And on it sits death…"

The Ramayana

"Dense arrows of flame,

like a great shower,

issued forth upon creation, encompassing the enemy…

A thick gloom swiftly settled upon the Pandava hosts.

All points of the compass were lost in darkness.

Fierce winds began to blow.

Clouds roared upward,

showering dust and gravel.

Birds croaked madly…

The very elements seemed disturbed.

The sun seemed to waver in the heavens.

The earth shook,

scorched by the terrible violent heat of this weapon.

Elephants burst into flame

and ran to and fro in a frenzy…

over a vast area,

other animals crumpled to the ground and died.

From all points of the compass

the arrows of flame rained continuously and fiercely."

The Mahabharata

In the 1950's when archaeologists were excavating the cities of Harappa and Mohenjo Daro, they discovered many skeletons of people lying in the streets, some holding hands, lying just as they had fallen from some horrible instantaneous event that annihilated them.

David Hatcher Childress states in his book, *Vimana Aircraft of Ancient India & Atlantis*, the following:

"These skeletons are among the most radioactive ever found, on a par with those at Nagasaki and Hiroshima. At another site in India, Soviet scholars found a skeleton with a radioactivity level in excess of fifty times that which is normal.

Other cities have been found in northern India that indicate explosions of great magnitude. A city was found between the Ganges and the mountains of Rajmahal which seems to have been subjected to intense heat. Huge masses of walls and the foundation of an ancient city were found fused together, literally vitrified!"[9]

The following is taken from Childress's book: "....Dr. Oppenhiemer, the "Father of the H-Bomb," was also a Sanskrit scholar. Once when speaking of the first atomic test, he quoted the *Mahabharata* saying, "I have unleashed the power of the Universe. Now I have become the destroyer of worlds." Asked at an interview at Rochester University seven years after the Alamagordo nuclear test whether that was the first atomic bomb to ever be detonated, his reply was: "Well, yes," and added quickly, "in modern history."[10]

A test of a weapon that resulted in similar effects was conducted above ground on an atoll of the South Pacific after World War Two. The occupants of the atolls in the area were evacuated. The bomb was set up in the center of the land mass. Into the beach the Navy drove a great post onto which they affixed a massive steel cable which stretched 5 miles into the sea. At regular intervals on the cable a barge was tethered and populated with live pigs. The physiology of pigs is similar to humans. The Navy anchored several miles away, just in sight of the trees of the atoll. When the bomb was detonated witnesses on the vessel reported they could hear an ear-piercing scream as thousands of pigs perished in the stream of high-energy neutrons killing them. Close to ground zero they perished instantly. Further out, they died slowly and painfully. Much was learned about how nuclear radiation kills in inverse proportion to the distance from the nuclear event.

About 14,000 B.C., Atlantis attempted to subjugate a civilization in the area of the Gobi Desert, called the Ulger Empire. They developed a technology by harnessing the Life Force with the use of giant crystals, called, *firestones*, to create "Scalar Wave Weaponry,"[11] which was fired through the center of the earth to annihilate their adversaries. In doing so, by the misuse of this power they nearly did themselves in; which misuse caused a second part of their island continent to sink into the depths of the ocean taking many lives with it. This second cataclysm resulted in a mass exodus or invasion of people onto the continent, spoken of earlier in this chapter.

Another name for the *Firestone*, commonly used by the Atlanteans, was the term *"Tuaoi stone,"* to describe their power source. The stone was in the form of a six-sided figure and set as a crystal. The clairvoyant Edgar Cayce, in reading no. 2072-10, July 22, 1940, states, "In the beginning, it (the tuaoi stone) was the source from which there was the spiritual and mental contact." It would seem that this is a reference to the "seer stones," that Nephilim possessed.

The Israelites possessed similar stones referred to in *Exodus* 28:30 as the Urim and Thummim. Could it be that the Urim and Thummim and the *Tuaoi Stone* are one and the same? Perhaps there were several sets of these stones. If one could locate a Urim and Thummim and apply the Life Force energy (electromagnetism) to either of them, would the energy activate these stones into becoming a power station such as was used by the Atlanteans? This power source was also used for medical treatment and for the rejuvenation of bodies through the application of the rays from the stone. Atlantis had several of these power stations that were unintentionally "tuned too high," which misuse caused the second destruction of the land.

The Naacals from Mu sent colonizers to what is known today as North Africa, and the Mediterranean Basin or Egypt. This was the Osiris Civilization. *Ra* was the first ruler of ancient Egypt, whose symbol, was the golden winged sun disc.

The Winged Disc

Like *The Empire of the Sun* and *The Rama Empire,* Egypt worshipped a solar deity.

After making an extensive study of all available writings and texts on *The Empire of the Sun,* the Atlantis civilization, the Osiris civilization and *The Rama Empire,* it is the opinion of the authors from the evidence found that the emperors Ramu, Rama, and Ra, were great rulers of these empires. They appear to have been immortal and divine beings, but nevertheless appear to be fallen sons (Elder Gods or Nephilim) of the Supreme God.

The Hindu's however, share a different view. Besides being a great warrior and emperor, Rama in Hinduism was an incarnation of Vishnu the Preserver, one of the trinity of supreme gods. Most Hindus in India believe Rama to be an actual historical figure. His exploits and adventures in the *Ramayana* have made him a sort of national hero in India. In the words of Murli Manohar Joshi, "Rama is not a Hindu god but a national hero, and every Indian irrespective of his religion must accept that." Joshi made this statement in an article which ran in *Time* magazine, November 12, 1990, when militant Hindus were marching on a town to reclaim a site that they believe to be Rama's birthplace, now occupied by a dilapidated Muslim mosque. The Hindus had hopes of building a new shrine to Rama on the old mosque site. Joshi was a supporter of this movement.[12]

Another known colony was in Lhasa, Tibet. These were centers of learning where maps, charters, and records of the history of the world and mankind were stored in repositories and libraries for safekeeping.

David Hatcher Childress's book, *Vimana Aircraft of Ancient India & Atlantis*, makes for fascinating reading. He writes the following, "The *Ramayana* describes a vimana as a double-deck, circular (cylindrical) aircraft with portholes and a dome. It flew with the "speed of the wind" and gave forth a "melodious sound" (a humming noise?). "The ancient Indians themselves wrote entire flight manuals on the control of various types of Vimanas, of which there were basically four: the Shakuna Vimana, the Sundara Vimana, the

Rukma Vimana and the Tripura Vimana. He goes on to say that, "the Tripura Vimana was powered by motive power generated by solar rays and had an elongated form similar to a modern blimp."[13]

Reading further, Childress quotes from an ancient Indian text called the *Samarangana Sutradhara*, as quoted by Ivan T. Sanderson, and Desmond Leslie, the following:

"Strong and durable must the body be made, like a great flying bird, or light material. Inside it one must place the *Mercury-engine* with its iron heating apparatus beneath. By means of the *power latent in the mercury* which sets the driving *whirlwind* in motion, a man sitting inside may travel to great distance in the sky in a most marvelous manner.

Similarly by using the prescribed processes one can build a vimana as large as the temple of the God-in-motion. Four strong *mercury* containers must be built into the interior structure. When these have been heated by controlled fire from iron containers, the vimana develops thunder-power through the mercury. And at once it becomes a pearl in the sky.

Moreover, if this iron engine with properly welded joints be filled with mercury, and the fire be conducted to the upper part it develops power with the roar of a lion."

We now call these *Mercury Vortex Engines.*[14]

Childress further states the following, "Recently, Sanskrit documents discovered by the Chinese in Lhasa were sent to India to be studied by experts there. Dr. Ruth Reyna of the University of

Chandigarh said that the manuscripts contain directions for building interplanetary spaceships!

Indian scientists were at first extremely reserved about the value of these documents, but became less so when the Chinese announced that certain parts of the data were being studied for inclusion in their space program! These airships, perhaps powered by "anti-gravity," as the document found by the Chinese suggests, are theoretically kept in secret bases. These airships, virtual UFO's, may actually be the cause of some UFO sightings, especially those in Central Asia."[15]

A single drop of mercury excited with a jolt of electricity and held in that state with a smaller amount of electrical power will form a plasma that emits ultraviolet energy 254 nanometers in wavelength. This energy, exciting a powdered dope inside a glass tube, comprises the modern fluorescent light bulb and may operate for ten years or more. Its pure radiation can generate chemical free radicals used in curing polymers or in sterilizing surfaces of all germs and viruses.

The following is taken from Judd's *Atlantis-Mother of Empires*: "According to the records held by the Asiatic Society of Calcutta, India, of about 100 years ago, the ancient Hindu books frequently refer to the lost Atlantis as The White Island, or the White Devil. In volume eight of the *Asiatic Researches*, page 286, we read: "This Atlantis was overwhelmed by a Flood." Also, in this remarkable collection of Hindu records, we learn that the early Egyptians conquered and ruled India."[16]

Continuing from the same book: Heinrich Schliemann, the discoverer of Troy and Mycene in 1873, states "there is a papyrus roll, written in the reign of Pharaoh, relating how that monarch sent an expedition *to the west* in search of traces of *The Land of Atlantis,* whence, 3,350 years before, the ancestors of the Egyptians arrived, "carrying with themselves all the wisdom of their native land." Pharaoh Sent or Senta, lived in the Second Dynasty, or about 4,100 B.C., although he is mentioned in connection with a writing of the First Dynasty. This date, added to 3,350, places the date of the arrival of the ancestors King Sent refers to as 7,450 B.C."[17] It is also interesting to note that the words *Atlas* and *Atlantic* have no satisfactory etymology in any language known to Europe. They are not Greek, and cannot be referred to any known language of the Old World. The root of the words is not known.

The Nephilim, having come from other dimensions, had much knowledge of the workings of the great universes. They understood the laws and physics of the temporal universe, therefore they were able to harness and utilize various types of forces in technology. When this is understood, then the idea of having airships, interplanetary spaceships, laser weapons and trans-oceanic flights to different parts of the world, many thousands of years ago, certainly has more credence. Picture the Nephilim as the *Darth Vaders* of our universe, using star war technology. Since their eviction from heaven, these spiritual beings have been battling for control (dominion) of space and planets, in this universe, to call home. It is the age-old story of the "sons of darkness" against the "sons of light" that has been immortalized in fables, myths, and even sacred scriptures.

According to Plato, in 9,600 B.C., the remaining portion of Atlantis sunk into the ocean after another cataclysmic event. This would be the cataclysm that the authors Allan & Delair in their book, *When the Earth Nearly Died*, state occurred in 9,500 B.C. Amazingly, these dates are separated by only 100 years!

The following account is called **The Rattling of the Pleiades**, which no doubt refers to the same cataclysm of 9,600/9,500 B.C.

The *Chilam Balam of Chumayel* speaks of an almighty cataclysm during which the "Great Serpent" was ravished from the heavens together with the rattles of its tail so that its skin and pieces of its bones "fell here upon the earth." The Calina Carib tribe in Surinam (formerly Dutch Guiana), South America, still have myths of a fiery serpent that came from the Pleiades and "brought the world to an end with a great fire and a deluge."

The Elder Gods, having been forewarned, fled this planet and watched the destruction from afar. *The Zohar* states that they trembled with fear in the heavens as they witnessed the horrible sight. Ninety percent of all life perished on this planet. We want to impress upon the reader that the Great War that started in heaven or the highest spiritual universe continues on in the temporal universe. It is good battling against evil and still is being played out on this planet and other worlds. Lucifer remains the source of most of the problems and evils that plague mankind.

Most of us don't realize that we continue to pay tribute to these gods on a daily basis, and have been doing so for over two thousand years! The very days of the week are named after them as follows:

Monday is named after the moon goddess, known as Selene, Luna, or Mani.

Tuesday is named after the Roman god of war, Mars.

Wednesday is named after the Roman messenger god, Mercury.

Thursday is named after the Roman god of thunder, Jupiter.

Friday is named after the Norse god Odin's wife, the goddess Frigga.

Saturday is named after the Roman god Saturn.

Sunday is named after the Roman god of the sun, Sol.

The names of the seven days of the week were adopted in Rome about 400 A.D., and spread into Europe, but were in use long before that in the East. The Hebrews refused to pay them tribute preferring instead to simply call each day of the week by a number, Sunday being their 1st day of the week etc.

We pay tribute by naming some of the months of the year after them as follows:

437

January was named after the god Janus, the Roman god of gates and doorways.

March was named for the Roman god of war, known to the Greeks as Ares.

April was named for the Roman goddess, Venus, known to the Greeks as Aphrodite.

May was named for the Italic goddess of spring, Maia.

June was named for the goddess of marriage, Juno. She was the sister of the god Jupiter.

One of the most sacred Christian holidays, Easter, is named after the ancient Olympian goddess of spring, Ishtar or Eshdar, but known to the Romans as Venus, to the Greeks as Aphrodite, to the Akkadians and the Sumerians as Inanna. To the Babylonians she was Astarte, the wife of Baal, the Sun God. The word *Easter* was added to the *New Testament* in one of many translations and does not exist in the ancient texts. The symbols of rabbits, eggs, and the custom of hunting for hidden eggs are carry-overs from the pagan rites of Spring.

Celebrating Christmas has been controversial since its inception. Any historian knows that taxes were rendered from the firstlings of the flock and field and payment was due the empire in the early Spring. Jesus was born early in the fourth month of the year, not in December. The Romans celebrated Saturnalia, a festival dedicated to Saturn, the god of peace and plenty in December. The council led by Constantine combined their Christmas holidays with the Roman holidays of Saturnalia as a masterful plan to more easily establish Christianity as the state religion among the heathen nations. From the festival of Saturnalia came the lights, merriment, feasting, frequent excesses and gift giving that we see today. This part of Christmas had its roots in pagan practices honoring the god Saturn who was one of the Titans or Nephilim! The simplicity of the Nativity was meant to be a holy day without the contrast of merriment and excesses that we see today.

Another popular ritual is the burning of the Yule Log, which is strongly embedded in the worship of the gods of vegetation and fire. The use of mistletoe and holly is associated with the god of fertility. Throughout the centuries and into the present, many people have condemned such practices as being pagan and contrary to the true spirit of Christmas. Modern efforts to commercialize the celebration for all holiday shoppers have nearly sterilized the season of any mention of Jesus. It culminates with Santa delivering toys to children, not with the birth of the Savior of mankind. Forty percent of all retail profits are realized between Thanksgiving and Christmas. Traditional American Thanksgiving tables often display a horn of plenty, sometimes woven like a basket. This horn, called a cornucopia, is actually a horned fertility symbol that was a tribute to Baphonet's fertility and symbolized Zeus being suckled by a goat whose horn broke off and was magically filled with fruit.

Baphonet was a pagan fertility god associated with creative force of reproduction and had the head of a goat.[18]

And if the above are not enough to satisfy our pagan appetites, even the planets, in our solar system, bear their names. We have associated them with our sun, moon, planet Earth, astrology, and even named some stars, satellites, and automobiles in their honor! Have we lost our minds or what are we thinking?

Of course, we have to mention the clairvoyant Edgar Cayce's prediction, in 1940, when he said, "And Poseidia will be among the first portions of Atlantis to rise again. Expect it in 1968 or 1969; not so far away!"[19] He further predicted, "...for these mountain tops (the Bimini Islands)-especially that along the north and eastern shores of the north and northern portion of the south island—will produce many various minerals and various other conditions that will be remunerative when the projects are undertaken."[20]

Explorations on Bimini, during the 1960's, 70's, and 80's have produced inconclusive results. The famous Bimini road site provided more questions than answers. These sites recently have had a renewed interest in exploration using high technology approaches to solve the mystery of Bimini and with luck rediscover Atlantis.

Did the lost civilizations of Mu and Atlantis ever exist? If you had asked us that question ten years ago, the answer would be probably not. Only a few historians or archeologists even mentioned it before 1985 for fear of being cast out of the mainstream of their professions. After having completed a comprehensive compilation of data on the Fallen Angels, from both ancient and historical worldwide perspectives, a different story began to emerge, one that we have tried to present in this book. Uncovering the fact that the Greeks also referred to the Titans as Elder Gods, Els, and Elanakim, short for Elder Race or simply the Giants, was a major turning point that made us realize the Greeks knew this race of gods to be actual historical figures that once lived on the fabulous island of Atlantis and elsewhere in the world. Their myths and fables, we now believe, are based on actual historical events. In the past, stories of the Giants were fragmented and the overall picture unclear. When we complied and reconciled all our researched data in a timeline format on the Giants, it quickly became apparent that this was a major piece of the Earth's history that has not been accounted for or at least taken seriously.

If these civilizations once existed then where is the evidence? The megalithic stone cities and gigantic stone structures found worldwide are part of the remaining evidence. The builders possessed technology that we today cannot duplicate, and yet various scientific groups persist in calling them a primitive Stone Age people. The remnants of the lost island "continent" of Mu, as tradition persists among the pacific islanders of today, are the volcanic mountaintops in the South Pacific and possibly Australia. Although the ruins of Atlantis have not yet been located, the sudden and abrupt western immigration patterns onto the European and African continent of an already civilized people, with advanced technology, strongly indicates Atlantis was the source. The cataclysm may have destroyed all traces of the island through tectonic plate upheavals and crust displacement, however, it appears that the myths and fables associated with the gods have actual historical basis. One more thought, the gaping hole left in the earth's side after the cataclysm may have been the former site of Atlantis! If so, then the giant statue of Poseidon may be in orbit somewhere in outer space.

Again, if you ask us the same question today, "Did the lost civilizations of Mu and Atlantis ever exist?" The answer is yes, we believe that they most likely did! Still a skeptic? Read the next chapter.

End Notes

[1] *(Atlantis Blueprint, Colin Wilson and Rand Flem-ath, Delacorte Press, N.Y., 2000, p. 4)*

[2] (Robinson p. 60)

[3] (Bordes, Francois, *"The Old Stone Age,"* McGraw-Hill Book Co., N.Y., 1968; Clark, J.Desmond, *"The Prehistory of Africa,"* Praeger University Series, N.Y., 1970; Coon, Carleton s., *"The Story of Man,"* Alfred A. Knopf, N.Y., 1954)

[4] . (Spence, Lewis T., *"The History of Atlantis,"* Rider & Co., London, 1926)

[5] (Smith, Phillip E.L., *"Stone Age Man on the Nile,"* Scientific American, Vol. 235, No. 2, August 1976.)

[6] (Hadingham, Evan, *"Secrets of the Ice Age,"* Walker and Company, N.Y., 1979)

[7] (Ref; Spence, Lewis T., *"The History of Atlantis,"* Rider & Co., London, 1926; Thorndike, Joseph J.Jr., *"Mysteries of the Past,"* American Heritage Publ., N.Y. 1977.)

[8] *Vimana Aircraft of Ancient India & Atlantis,* ©1991. Published by Adventures Unlimited Press, Kempton, ILL., pp.59-62)

[9] *Vimana Aircraft of Ancient India & Atlantis,* ©1991. Published by Adventures Unlimited Press, Kempton, ILL. 1991, pp.81-82)

[10] (*Vimana Aircraft of Ancient India & Atlantic,* et al p. 83)

[11] (*Vimana Aircraft of Ancient India & Atlantis* et al p. 89)

[12] (*Vimana Aircraft of Ancient India & Atlantis,* et al pp.308-309)

[13] (*Vimana Aircraft of Ancient India & Atlantis* et al p. 36)

[14] (*Vimana Aircraft of Ancient India & Atlantis,* et al p. 254)

[15] (*Vimana Aircraft of Ancient India & Atlantis* et al p. 26-27)

[16] *Atlantis Mother of Empires,* ©1939 by Robert B. Stacy-Judd. Published by Adventures Unlimited Press, Kempton, ILL. p. 90)

[17] *Atlantis Mother of Empires,* ©1939 by Robert B. Stacy-Judd. Published by Adventures Unlimited Press, Kempton, ILL. p. 84)

[18] (*The Da Vinci Code* copyright Doubleday 2003 p 316-317)

[19] (reading no. 958-3)

[20] (reading no. 996-12, March 2, 1927)

E. J. Clark & B. Alexander Agnew, PhD

The Brave New World

There were survivors of the 9,600/9,500 B.C., cataclysm, however, the topography, time, and climate had all changed. Previous to the cataclysm was a period often referred to as the Golden Age. The earth then had a near vertical axis. This axis made practically all of earth's land mass temperate, almost tropical, the exception was a few areas in the polar regions where it might have been cold enough for some icing. Hot and cold climate temperatures were unknown and day and night were equally divided, however they may have been longer due to a slower rotation of the earth, perhaps thirty hours opposed to twenty-four of today. Biological evidence shows a remarkable proliferation of plant and animal life. Vegetation was richer and more abundant than that of today. Mountains were lower and seas shallower. Earth was a virtual paradise. Allan and Delair state in their book, *When the Earth Nearly Died,* the following:

"Studies have shown that at 10, 178 B.C., or over 12,000 years ago, the celestial pole was inclined at an angle of 30 degrees from its present position. This in turn strongly suggests that the terrestrial axis was then oriented differently from today. It also suggests that at the time of the Deluge (cataclysm) which allegedly terminated the Golden Age there must have been a change to the tilt of the Earth's axis.

Any axial changes of the type just suggested would, of course, have been attended by various other major natural adjustments on quite literally a global basis.

It is also clear that the loss of this ancestral paradise involved a rearrangement of the seasons. Indeed, describing the "World Before the Present," the Navajo Indians of North America insist that:

"...The seasons were much shorter than they are now. A year then was but a month of our time."[1]

The Elder Gods or Nephilim were fore warned by their "seer stones" of the impending disaster coming from an errant interloper on collision course with planet earth. They foresaw the event by nearly 1,000 years of their time then. They immediately took measures to preserve some of the earth's population from complete annihilation. Underground cities were dug out in various parts of the world to save part of the population. Besides having the ability to "see" into the future, they could discern what areas were the most survivable. In 1964, in eastern Turkey near Kaymakli, such an underground tenement was discovered with several levels and corridors 10 feet wide and 6. 5 feet high. Huge round stones, similar to millstones, that could be 'locked' from the inside, closed doorways. Beneath the town of Derinkuyu, another underground city was discovered that had eight stories and was big enough to hold a population of 20,000. This underground tenement was completely invisible from the ground, in spite of ventilation shafts. These dug out cities are believed to date back before 10,000 B.C.[2] These are two of the 36 and possible 40 underground cities of Cappadocia. Not all have been explored due to cave-ins. A few are open to the public.

Another colonizing branch of the Atlanteans arrived in North Africa prior to the cataclysm and populated the Triton Sea's rim and some of its islands, referred to in *The Histories of Herodotus*. One large island had centrally located low- lying mountains shaped like a trident, the symbol of Neptune's Atlantis, their homeland. This colony of Atlanteans called themselves the *Tuaraks*. They hollowed out these mountains to establish libraries for their sacred texts, a safe haven, and place for their sacred temple rites.

Another range of mountains located in their rim territories were the *Atlas Mountains,* named after Atlas, one of the sons of Neptune. It is said that the giant *Hercules*, son of the Olympian god, Zeus (Jupiter) and Alcmente, helped the Tuaraks build a second set of underground tenements in this mountain range. No doubt, the Tuaraks rode out the cataclysm in these underground safe havens, having been fore warned by the gods. After the cataclysm that sunk Atlantis, the Triton Sea was drained and their island low lying mountain range had risen to towering mountains which became part of the Ahaggar Mountain range of southern Algeria. In time the area around North Africa became an arid desert known as the Sahara. The descendants of the Tuaraks are known today as the *Tuaregs*, who are the nomadic Berber traders of the Sahara. Tourists are permitted today to visit the underground caverns of the Ahaggar Mountains and experience for themselves the intact remains of an

Atlantean civilization. Visitors are even allowed to tour the Tuareg under ground library whose shelves are packed with thousands of ancient texts, some dating back to the time of Atlantis.[3]

Intricate tunnel systems with cavernous crypts connected the major Andean cities and temple sites, previously discussed in another chapter. These tunnel systems ran hundreds of miles through Peru, Ecuador, and Bolivia, some which have been partly explored. Legends speak of a Pan-American system of tunnels that once connected all the Americas, terminating in the Arizona desert.[4] Besides being used as an underground highway, the tunnels afforded protection and safe havens for their treasures, maps, and sacred written records.

In Erich Von Daniken's book, *The Gold of The Gods,* Von Daniken and archaeologist Jaun Moricz, an Argentine subject, explored a hidden tunnel system in Ecuador in 1969, which is guarded by hostile Indians. The tunnel system is situated within the triangle formed by the three towns Gualaquiza, San Antonio, and Yaupi, which are located in the province of Morona-Santiago. Von Daniken states in his book, "My doubts about the existence of the underground tunnels vanished as if by magic and I felt tremendously happy. Moricz said that passages like those through which we were going extended for hundred of miles under the soil of Ecuador and Peru." As they were exploring the tunnels, they entered into a gigantic hall full of statues of animals and the most precious treasure of all, a metal library! Von Daniken writes, "The library of metal plaques was opposite the zoo, to the left of the conference table. It consisted partly of actual plaques and partly of metal leaves only millimeters thick. Most of them measured about 3 feet 2 inches by 1 foot 7 inches. After a long and critical examination, I still could not make out what material had been used in their manufacture. It must have been unusual, for the leaves stood upright without buckling, in spite of their size and thinness. They were placed next to each other like bound pages of giant folios. Each leaf had writing on it, stamped and printed regularly as if by a machine. So far Moricz has not managed to count the pages of his metal library, but I accept his estimate that there might be two or three thousand." Von Daniken further states, " that the characters on the metal plaques are unknown." Moricz felt these metal plaques might contain a resume of the history of a lost civilization.

Moricz later went and claimed a title deed on the caves beneath Ecuador on July 21, 1969, and put them under state control which will smooth the way for future research. The title deed makes him the legal owner of the metal library and other objects in accordance with Article 665 of the Civil Code. His request for an audience with the President of Ecuador was met with long delays that apparently never materialized. It appears that President Velasco Ibarra had no time to give to an unknown archaeologist with tales of an incredible discovery.[5] Apparently, these tunnels do exist. In 1973, a British group of explorers partially explored some of the tunnels. No mention was made of any artifacts found.

Chinese and Tibetan legends contend that most all ancient Asian cities were once connected by a network of underground passages and caverns which ran under the entire continent. Colonizers from Mu were said to have taken protection from certain destruction in these subterranean caves from an impending disaster and deluge. Some of these tunnels terminate in Tibetan monasteries, the cave temples of Tunhung in western China, and Ellora and Ajanta in India.[6]

The Egyptians had a vast network of underground tunnels and caverns that connected the pyramids with their temples. In addition they had numerous catacombs and subterranean crypts of vast extent in Thebes and Memphis that were used as safe havens, burials, and sacred rites. It is now known that the entire Giza plateau is honeycombed with underground tunnels. It is conceivable that the Sphinx was constructed hurriedly by the gods before the cataclysm to mark the site of the Hall of Records. Knowledge was important to the Elder Gods and it was urgent to preserve their records for the future in a safe place. It is said that the god, Thoth-Hermes of Atlantis, recorded vital information on emerald tablets, and deposited them in the Hall of Records before the Deluge. The Giza site marks a powerful vortex and therefore would be easy to find should the Sphinx fail to survive the impending disaster.

Some survived by being on high ground or mountaintops. Most all claimed to have had divine warning in advance by the gods. The following is an extract from *When the Earth Nearly Died:*

"In the well-known Greek story of *Philemon and Baucis,* for example, the righteous Phrygian couple were taken by the gods Zeus and Hermes to the top of a lofty mountain to escape the fate reserved for their fellow men, who perished shortly afterwards in a Deluge."[7]

The Chinese relate a similar story of Maligasima, that sounds a whole lot like Atlantis, as follows,

"The continent which legend relates formally sank beneath the ocean's waves as related in a legend owing to the iniquity of the giants in Ma-li-ga-si-ma, it was submerged with all its inhabitants, except the King, Periru-un, who was able to escape from the deluge with his family, having been warned by the gods of the impending disaster (catastrophe) through two idols. This King and his descendants peopled China."[8]

From Judd's *Atlantis-Mother of Empires*: The Hindu *Bhagavata-Purana* tells us that the Fish-God, who warned Satyravata of the coming of the Flood, directed him *"to place the Sacred Scriptures in a safe place."* In Berosus' version of the Chaldean Flood, he says: "The deity Chronos (elder god) ...warned him (Xisuthhros) that...there would be a flood by which mankind would be destroyed. He therefore enjoined him to *write a history of the beginning, procedure, and conclusion of all things, and bury it.*"[9]

Others survived by building boats (arks) in advance by divine warnings, others not listening to warnings grabbed what ever floating objects they could and hung on for dear life, these being the true Deluge heroes in many legends.

World wide are legends of refugees and animals becoming entombed in caves with their entrances blocked with mud, boulders, and rocks borne by the flood waters. Survivors had to literally dig their way out to the surface. It is obvious that deep caverns would not afford protection from rampaging flood waters, unless located in high mountain places, even then most were death traps. Of the shockingly low number of humans who survived, almost all lived in the higher altitudes of mountain ranges or purposely took shelter there in caves.

The cataclysm of 9,600/500 B.C., was known traditionally as the Flood of Deucalion. The flood of Noah was called the Flood of Ogyges.[10] On the temporal earth, the Flood of Deucalion occurred before the Flood of Ogyges, however, they really occurred at the same time. Remember the spirit earth was in another time and dimension when the Flood of Ogyges occurred. Both creations were experiencing destruction's at the same time, but one creation was in another dimension. Allan and Delair state, "Commyns Beaumont (Appian Way) is emphatic on this point:.... Tatian, Clement and Eusebius all agreed that the Phaeton event (cataclysm) was identical with the Deucalion flood."[11] Now your authors think differently on the names of the two floods. We believe that Tation, Clement and Eusebius didn't know about the union of the two earths and have confused, as many have done, the cataclysm with the flood of Noah. We believe that the Egyptian Gnostics had it correct with the following statement found in *The Apocalypse of Adam (v, 5)* "And God will say to Noah—whom the generations will call Deucalion".

It now appears that the cataclysm was the Flood of Ogyges and the flood of Noah was the Flood of Deucalion. We felt it necessary to present both documented views because scholars of the above texts would have called us to task. The names of the two floods and when they occurred have long confused writers, both ancient and modern.

Allan and Delair further state, " the Earth's rate of spin was probably the single most important factor which determined its survival as a planet.... Earth's slower rotational rate prevented it from breaking up—although not, in our submission, by any great margin.... . Earth's survival apparently came close to a situation best described as 'touch and go.' "[12]

Legends worldwide speak of survivors, but the real test of survival was yet to come. Following the catastrophe the earth was plunged into darkness ensued with unbearable cold that endured for years after the floodwaters had subsided. This period has been traditionally remembered as the *Age of Darkness*. Numerous accounts speak of a 'collapsed sky' condition where people could not stand upright without literally touching the low firmament. Several African tribes share a belief of ancient times when the god Kagra threw the firmament upon the earth to annihilate humanity.[13]

The following is from *Huai nan zi* (writings of Chinese Prince Huainan)

Nu Wa Mends the Sky

"In ancient times, the Four Corners of the sky collapsed and the world with its nine regions split open. The sky could not cover all the things under it, nor could the earth carry all the things on it. A great fire raged and would not die out; a fierce flood raced about and could not be checked. Savage beasts devoured innocent people; vicious birds preyed on the weak and old.

Then Nu Wa melted rocks of five colours and used them to mend the cracks in the sky. She supported the Four Corners of the sky with the legs she had cut off from a giant turtle. She killed the black dragon to save the people of Jizhou (the central one of the nine regions), and blocked the flood with the ashes of reeds. Thus the sky was mended, its Four Corners lifted, the flood tamed, Jizhou pacified, and harmful birds and beasts killed, and the innocent people were able to live on the square earth under the dome of the sky. It was a time when birds, beasts, insects and snakes no longer used their claws or teeth or poisonous stings, for they did not want to catch or eat weaker things.

Nu Wa's deeds benefited the heavens above and the earth below. Her name was remembered by later generations and her light shone on every creation. Now she was traveling on a thunder-chariot drawn by a two-winged dragon and two green hornless dragons, with auspicious objects in her hands and a special mattress underneath, surrounded by golden clouds, a white dragon leading the way and a flying snake following behind. Floating freely over the clouds, she took ghosts and gods to the ninth heaven and had an audience with the Heavenly Emperor at Lin Men (the place where the heavenly gods lived), where she rested in peace and dignity under the emperor. She never boasted of her achievements, nor did she try to win any renown; she wanted to conceal her virtues, in line with the ways of the universe."

It is impossible to calculate just how many beings, including perhaps some giant offspring, survived the actual event of the cataclysm but probably more than half of the survivors died from exposure to the terrible cold without proper clothing, lack of fresh food, and in general, hellish living conditions. Men were reduced to living like beasts, having to eat the frozen flesh of dead animal carcasses, some probably practiced cannibalism, a practice that remained prevalent, particularly in South America and the South Pacific Islands, until the early part of the last century. Only the strongest managed to survive. At that time, the future of mankind was unquestionably "hanging in the balance." Humankind had been literally catapulted into a new age of uncertainty. It was adapt or perish. It was the end of the Golden Age, First Time or Zep Tepi as called by the ancient Egyptians, and the Fourth Sun.

Another excerpt from *When the Earth Nearly Died* states, "The ancient association of a period of intense cold on Earth with some celestial influence is also found in the *Old Testament*, which links the icy phase with the Pleiades and with *Mazzaroth* (alias Phaeton/Marduk) for we read:

"Out of whose womb came forth the ice?
And the hoary frost of heaven,

who hath gendered it?

The waters are hid as with a stone

and the face of the deep is frozen.

Canst thou bind the sweet influences of Pleiades,

or loose the bands of Orion?

Canst thou bring forth Mazzaroth in his season?"[14]

In spite of overwhelming circumstances, mankind did adapt in a most marvelous way. They banded together in tribes, pooling and utilizing their talents and resources. They migrated to warmer latitudes, where the collapsed sky condition was less pronounced, and to areas where the landscape had proceeded with re-vegetation. DNA studies have shown that around the time of the cataclysm, there was a bottleneck effect, meaning a sudden catastrophic decrease in population, and shortly afterwards influx of migration into different parts of the world.

Indeed, it was a different world. The topography of the entire globe had changed. The new inclined axis of the now wobbling unstable earth, caused severe weather conditions. Coastlines had changed, tall mountains had risen, many land areas had sunk during fracturing of the earth's crust, others had risen, oceans were higher, some seas drained and others formed. One particular landmass that sunk was the Mediterranean basin, formerly an elevated valley consisting of lakes prior to its disruption and submergence. The Strait of Gibraltar did not exist at that time.[15] Eventually, after a long period of time, the collapsed sky conditions returned to normal. The combined conditions of enforced degraded tribal living, a wandering nomadic existence, and harsh planetary changes caused long- term retardation in mankind's progress for decades. Memories of the cataclysm were explained as a "great war in heaven" between the departed gods and eventually was preserved in many ancient myths as such.

Mankind existed and multiplied under these conditions for nearly 1,000 years. Then the Elder Gods returned to earth, some accounts say in "fiery flying machines," and re-established knowledge among men on a global basis. Once more they taught men how to read, write, construct cities, build ships, map making, agriculture, metallurgy, astronomy, math, medicine, divination, magic, the techniques of war, or in other words restore civilization. However, their main agenda was to re-corrupt the surviving seed of men to ensure that the redeemer would not be born here on this planet. Idolatry and priest-crafts were re-introduced under the auspices of religion. Once again, giants began to be born from the fair daughters of men. Legends speak of the bellies of women rupturing during childbirth because of the size of the giant infants. Even into the middle ages women feared childbirth because of the possibility of having giant offspring through ancestral lines being corrupted generations before. Married and pregnant women carried talismans to ward off this evil.

Different cultures and different times called the gods by various different names. The first name, given below, is the earlier name and the second name is the name known in later times to the same cultures. In Judaic tradition, they were called Nephilim or

Watchers; in Iranian legends, they were the Ahuras or Daevas; in Sumero-Akkadian legends, they were the Anannage or Annunnaki; and in Assyrio-Babylonian legends, they were the Edimmu. Later cultures of the Roman and Greek empires referred to them as Elder Gods or Titans. These were the immortals. Some of their gigantic off spring were given the status of immortality, but most were mortal, however, they usually enjoyed a long span of life. Most of the Olympians were immortal because they were the off spring of both male and female gods and goddesses, however some mated with mortal men and women and produced mortal Olympians. Kings who are descendent of partnerships between humans and the gods are considered either divine, or part-demon, or demi-gods.

The Elder Gods and their giant off spring, along with mankind, began to rebuild some megalithic cities, namely Tiahuanaco in Bolivia, and to construct new towns and megalithic stone cities and structures all over the world. By 8,000 B.C., the first new town, called, Jericho, was constructed. Other megalithic cities such as Ba'alBek and the original Jerusalem in the Middle East soon followed as did Babylon and Byblos. These cities were all built and occupied by giants. In fact, it is mentioned in *The Zohar* that King Nimrod was a giant. Readers are referred back to the chapter, *The Giants*, for a broad over view of their dispersal and occupation of this planet.

The ancient Egyptians claim they inherited their advanced culture from a race of Elder Gods who lived during the previous age known to them as Zep Tepi, or the First Time. They knew the

Elder Gods to be members of a race of fallen angels who founded ancient Egypt. Egypt apparently was spared the brunt of the cataclysm. The gods returned to the colonizers and helped them to build the Giza temple complex near the Sphinx. It is claimed that they built pyramids in the area, yet undiscovered, lying buried in the sands. Temple priests initially maintained the Hall of Records until schools of learning were built that housed copies of the texts and maps contained therein. Because of access to the Hall of Records that contained important knowledge vital to civilizations, Egypt quickly became the most advanced, learned, and powerful empire for thousands of years.

Erich Von Daniken states the following in his book, *Gods From Outer Space,*

"The Sumerians had one single concept for the universe: *an-ki*, which can be roughly translated as "heaven and earth." Their myths tell of "gods" who drove through the sky in barks and fire ships, descended from the stars, fertilized their ancestors, and then returned to the stars again. The Sumerian pantheon, the shrine of the gods, was animated by a group of beings who possessed fairly recognizable human shape, but appear to have been superhuman, indeed immortal. But the Sumerian texts do not refer to their "gods" with vague imprecision; they say quite clearly that the people had once seen them with their own eyes. Their sages were convinced that they had known the "gods" who completed the work of instruction. We can read in the Sumerian texts how everything happened. The gods gave them writing, they gave them instructions for making metal (heavenly metal) and taught them how to cultivate barley. We should also note that according to Sumerian records the first men are supposed to have resulted from the interbreeding of gods and the children of earth."[16]

As your authors have stated before, this is probably what the Anannage told men in order to hold them in willing submission and that is the reason for so many similar legends as to the origin of mankind as stated in *The Gods From Outer Space* as follows:

"Tibetan and Hindus called the universe the mother of the terrestrial race.

The natives of Malekula (New Hebrides) state that the first race of men consisted of descendants of the sons of heaven.

The Red Indians say that they are the descendants of the thunderbird.

The Incas believed they descended from the "sons of the Sun."

The Rapanui trace their origin back to the birdmen.

The Mayas are supposed to be children of the Pleiades.

The Teutons claim that their forefathers came with the flying *Wanen.*

The Indians believe that they descend from Indra, Ghurka or Bhima—all three of whom drove through the heavens in fire ships.

The South Sea islanders say they descend from the god of heaven, Tangalao, who came down from heaven in an enormous gleaming egg."[17]

Continuing on with Von Daniken, he further writes, "On one
of the tablets translated by the Sumerologist S. N. Kramer we read:
"In order to destroy the seed of mankind the decision of the council
of the gods is proclaimed. According to the commanding words of
An and Enhil.... **their dominion shall come to an end....**"[18]

It is in this text, by the "gods" own admission, their purpose
for coming to this planet was to corrupt the seed of men that will
end their dominion. It would end the dominion of this planet if the
redeemer couldn't be born and fulfill God's promise to the "upper
Adam." The redeemer had to come through uncorrupted pure
Adamic bloodlines.

The Sumerian *Epic of Gilgamesh* is even seen through
slanted "eyes." Gilgamesh was the 5th Sumerian King of Uruk.
He is described in the Epic as being 2/3 god and 1/3 human and is
also described as being a giant! *The Giants,* text of the *Dead Sea
Scrolls* mention Gilgamesh twice. In this epic the creation of man
is credited to the workings of the Anannage.... thus, as said before
those teachings kept men in willful submission.

Astonishingly the Sumerian King Records of ancient
Mesopotamia preserved in cuneiform script, called the Weld-
Blundell prism, gives accounts of rulers going back nearly half a
million years. Yet, even more amazing is that each king is said to
have ruled for twenty or thirty thousand years. King List WB 444
reveals: "When the kingship came down from the skies the kingship
was in Eridu. He reigned 28,000 years. Alagar reigned 36,000
years. Two kings each ruled for 64,000 years!" And so the list goes

on, one account after another of incredible reigns. After the deluge of 9,600/500 B.C., the reigns of the kings come down dramatically to just 15 years!

Chinese records refer to an age when "Heavenly Emperors" ruled here on Earth. Thirteen emperors are each said to have had reigns of 18,000 years. Here again their records insist this kingship arrived from the skies, a kingship that was to last for nearly half a million years. Almost every culture of size has oral traditions of when "sky rulers" reigned on this planet for thousands of years.

There have been many explanations for the fantastic number of years the kings each reigned. We believe that these records are further evidence of the Elder Gods occupation of this planet beginning many thousands of years ago. The kingship literally arrived here from the skies, established colonies, built cities, coexisted with men, and corrupted their bloodlines. They were immortal beings so that is why they could rule for thousands of years. After the 9,600/500 B.C., cataclysm mortals resumed rule of some of these ancient cities that were rebuilt and consequently their reigns were much shorter.

There is an interesting statement found in the Appendix section, under *History of the Jews, Book 5,* as found in the book *The New Complete Works of Josephus*, that reads as follows: " The Tradition is, that the Jews ran away from the island of Crete, and settled themselves on the coast of Libya, and this at the time when Saturn was driven out of his kingdom by the power of Jupiter."

Josephus continues, "But because their priests (the Jewish priests), when they play on the pipe and the timbrels, wear ivy round their head, and a golden vine had been found in their temple, some have thought that they worshiped our father Bacchus, the conqueror of the East; whereas the ceremonies of the Jews do not at all agree with those of Bacchus, for he appointed rites, that were of a jovial nature, and fit for festivals, while the practices of the Jews are absurd and sordid."[19]

In another section he writes, "And indeed Alexander Polyhistor gives his attestation to what I here say; who speaks thus: "Cleodemus the prophet, who was also called Malchus, who wrote a history of the Jews, in agreement with the history of Moses, their lawgiver, relates that there were many sons born to Abraham by Keturah: nay, he names three of them, Apher, and Surim, and Japhran. That from Surim was the land of Assyria denominated; and that from the other two (Apher and Japhran) the country of Africa took it name; because these men were reinforcements to **Hercules** (a giant son of Zeus or Jupiter), when he fought against Libya and Antaeus; and that Hercules married Aphra's (Apher's) daughter, and of her he begat a son, Diodorus; and that Sophon was his son, from whom that barbarous people called Sophacians were denominated."[20] Words in parenthesis and bold type added.

These writings imply that Saturn, Jupiter, Bacchus, and Hercules, were indeed historical figures that were recognized as gods and Bacchus was recognized as their father or creator. Hercules, a giant son of Zeus and Alcmene, apparently married Apher's daughter (Abraham's granddaughter) corrupting that branch of Abraham's offspring. After the cataclysm, the Elder Gods stayed only long enough to guarantee the corruption of the bloodlines and to re-establish civilization, then they departed because they hated the harsh living conditions in the new age, having preferred the paradise of the Golden Age. Life was simply too hard here and not to their liking. They reasoned this earth was just a little remote, insignificant planet, on the edge of the Milky Way, devastated by a cataclysm, the climate was now harsh with fierce storms, practically unfit for habitation, and all the seed of the survivors had been corrupted for generations. Before departing, they left their giant off spring in charge that reigned supreme for many generations.

And furthermore, they concluded that their mission here was complete, the redeemer would not come to this planet. They left to find other worlds to corrupt their seed in like manner to prevent the redeemer from being born on that particular planet. The billions of Nephilim were doing this to every planet they could find in the temporal universe that had human life forms in order to prevent the Redeemer being born on that creation. Tactics of this type, in fact if successful, would have guaranteed their dominion over planets in this universe in which to call home. Of course, this error in judgment resulted in the eventual coming of the redeemer to this planet. Why did he choose this little obscure planet? Some say the Redeemer came to this planet because it probably was the most wicked of all the inhabited planets. Have you thought there may be another, even more compelling logic? After all the efforts from those outside the family of Adam and Eve to corrupt and destroy His Plan, the evidence seems to indicate that He may have chosen Earth as the birthplace of the Savior because it was the **most unlikely place.**

The mystical union of the spirit earth with the temporal earth, the Flood of Deucalion, occurred in a quiet manner, without fanfare, unannounced, and unnoticed. It is rather difficult to pin point the exact date when this event occurred but we estimate that it may have occurred around 4,800 B.C. This date of course, will shock *Bible* scholars who place the date of the flood of Noah much later of about 2400-2300 B.C. However, they have lacked the knowledge of the union of the two earths, the knowledge of which dramatically clarifies the entire history of the earth.

Noah and his entourage arrived in an isolated region and remained in that area awhile before journeying to lower Sumer, later called "Chaldea", which occupied the same "Plains of Shinar." A branch of his family, some descendants of Ham, and his daughter, Egypt, migrated into Egyptian lands. In the book, *Atlantis-Mother of Empires,* by Robert B. Stacy- Judd, the following is stated: " There is actual historical evidence that a superior race of people invaded Egypt from the East, or Southeast, prior to 4,500 B.C. They brought with them a civilization superior to the African, and appear to have introduced wheat, barley, the sheep, the art of writing, a superior kind of brick-making, etc. They settled in the Nile Valley and Delta."[21] Delta regions are marsh lands full of reeds and rushes, such as was, and still is, a description of lower Mesopotamia and the regions near the Euphrates, which includes the Nile Valley. Such was the area that Egypt first discovered and later settled. From the East would have been from the direction of the Akkadian civilization, who later became known as the Chaldeans.

In a short time, after arriving on this earth, Noah and his people encountered another race of people not related in any way. What a surprise encounter that must have been! Believe it or not *The Zohar* describes such an encounter. The account goes as follows:

"They (two men) went on and sat down by a cleft in the rock, and were amazed to see a man suddenly emerge from it. 'Who art thou?' (the two men asked) 'I belong to the denizens of **Arqa**,' he answered. 'Are there human beings there?' they asked. 'Yes,' he answered, 'and they sow and reap. Some of them are of a strange appearance, different from my own; and the reason I ascended (came) to you is to learn from you the name of the earth wherein ye dwell.' 'This earth, is called **Erez**. (answered the two men). The man thereupon returned to his place, leaving them astonished.'"[22]

From *The Book of the Angel Rezial,* **Arqa/Areqa** is the name of the spirit earth that exists in the spiritual universe of Yetzirah, the universe of formation. According to the book, the lowest earth (the temporal earth) is named **Erez/Aretz,** located in the universe of making (the physical universe), called Assiyah. *The Book of the Angel Rezial* identifies the names of the two earths and the *Hebrew Book of the Dead* identifies the location of each earth.[23]

The unrelated people who had contact with Noah's family regarded them as divine gods similar to the Elder Gods. Like the Elder Gods, they had arrived in a great flying ship, the Ark of Millions of Years, which came from another place in the heavens, having sailed across the cosmic sea. To the children of this planet, they were indeed gods. No amount of explanation from Noah would have convinced them otherwise. They admired their great knowledge and longevity. Kings welcomed and bestowed honors, fit for gods, upon them. Pharaohs acknowledged their divinity, and thereafter referred to them as The Divine Seed or Holy Seed. As such, in the Babylonian Mysteries Noah was called "Dipheus" or the "twice born" god because he had lived in two worlds and was commemorated as a god with two heads looking in opposite directions, one toward each world.

It is generally accepted in Jewish tradition that Shem became King of Salem and was known in religious circles as the great high priest Melchizedek, however another account is found in *The Book of Enoch The Prophet,* as follows:

"Behold the wife of Nir—Noah's brother—whose name was Sopanim, being made sterile and never having at any time given birth to a child by Nir. Sopanim was in the time of her old age and in the day of her death. She conceived in her womb, but Nir the priest had not slept with her from the day that the Lord had appointed him to conduct the liturgy in front of the face of the people. When Sopanim saw her pregnancy she was ashamed and embarrassed, and she hid herself during all the days until she gave birth. Not one of the people knew about it. When 282 days had been completed, and the day of the birth began to approach, Nir remembered his wife. He called her to himself in his house so that he might converse with her. Sopanim came to Nir, her husband; and behold, she was pregnant, and the day appointed for giving birth was drawing near. Nir saw her and became very ashamed. He said to her, "What is this that you have done, O wife? Why have you disgraced me in front of the face of the people? Now, depart from me and go where you began the disgrace of your womb, so that I might not defile my hand on account of you, and sin in the face of the Lord.

And it came to pass, when Nir had spoken to his wife, Sopanim, that Sopanim fell down at Nir's feet and died. The archangel Gabriel appeared to Nir, and said to him, "Do not think that your wife Sopanim has died because of your error, but this child, which is to be born of her is a righteous fruit, and one whom I shall receive into paradise, so that you will not be the father of a gift of God.

When they had gone out toward the grave, a child came out from the dead Sopanim and sat on the bed at her side. Noah and Nir came in to bury Sopanim and they saw the child sitting beside the dead Sopanim, wiping his clothing. Noah and Nir were very terrified with a great fear, because the child was fully developed physically. He spoke with his lips and blessed the Lord.

Noah and Nir looked at him closely, saying, "This is from the Lord, my brother." And behold the badge of priesthood was on his chest, and it was glorious in appearance. Noah said to Nir, "Behold, God is renewing the priesthood from blood related to us, just as He pleases.

Noah and Nir hurried and washed the child. They dressed him in garments of the priesthood, and they gave him bread to eat, and he ate it. And they called him Melchizedek. The Lord heeded Nir and appeared to him in a night vision. And He said to him, "I, in a short while, will send my archangel Gabriel. And he will take the child and put him in the paradise of Eden. He will not perish along with those who must perish. As I have revealed it, Melchizedek will be My priest to all holy priests. I will sanctify him and will establish him so that he will be the head of all priests of the future.

And behold, Melchizedek will be the head of priests in another generation. I know that great confusion has come and in confusion this generation will come to an end, and everyone will perish, except that Noah, my brother, will be preserved for procreation. From his tribe there will arise numerous people, and Melchizedek will become the head of priests reigning over a royal people who serve You, O Lord."[24]

As you can see, Melchizedek was born of a virgin mother. He was taken from the earth and placed in the protective custody of the Garden of Eden until 11 years after the flood. He was returned to the earth by the angel Gabriel and assigned to duty of guarding the tomb of Adam, who was reburied by Noah after the flood waters had subsided, as his body and that of Eve were carried upon the Ark so as not to be washed away in the flood. He maintains his status as high priest even unto the days of Abraham. He remained the King of Salem, and was described in the book of *Hebrews* as the one who bestowed the priesthood upon Jesus, *at another time.* We do not know what happened to him after the mission of Jesus was completed. We do know he was described in *Hebrews* as being without beginning of days or end of years.[25]

What language did Noah speak? According to the *Book of Moses* 6:5-6, he would have spoken the pure Adamic language of God, no longer extant. This language, in its pure form, was far superior to any known tongue in existence. It is possible that the Adamic language was an early form of Hebrew that eventually blended with the Ancient Akkadian language to later emerge as the Archiac Hebrew dialect. Both languages are dead. The Akkadian language was replaced by Aramaic and the – Archiac Hebrew language was revived in the 1880's as Modern Hebrew.

In the British Museum pamphlet "The Babylonian Legend of the Flood," the following statement is made: "Akkadian is a Semitic language cognate to Hebrew, Arabic, Assyrian and Babylonian. Sumerian on the other hand, is an unclassified language which has so far resisted all attempts to relate it to any known living or dead language." All nations who are descendants of Shem, son of

Noah, are the Semites, a modern designation taken from the Hebrew Scriptures.

An ancient Akkadian legend depicts their ancestor, a king named Xisuthus, sailing a ship to Armenia after the great flood (Deucalion), and his son, Zerban, becoming supreme king of the region. Old Chaldean texts also name their primogenitor as a mighty prince named Zerban. Berosus, the Chaldean astrologer and priest of the Babylonian temple of Belus, identified Zerban as a Titan, that is, an elder god.

If the Akkadians/Chaldeans were descendants of Zerban and their language is considered, at least in part, Semitic, then this would indicate that possibly a branch of Shem's descendants became corrupted with blood lines from Zerban or Zerban's descendants marrying "a fair daughter of men." Not all Chaldeans were descendants of Zerban as some Semitic people, for instance, moved and took up residence into the area and became known as Chaldean whose bloodlines remained "pure." Such was the case perhaps with Abraham. If King Nimrod, a descendant of Ham, were indeed a giant, then corruption of bloodlines among Noah's descendants was more prevalent than previously realized. After the final end of Sumerian power and civilization around 2,000 B.C., the region came under exclusive control of Semitic peoples for centuries.

Now there were two races of people upon planet earth. One race was the pure Adamic Race or Race of Adam and the other was the Hu'man Race. Noah and his family were the descendants of the pure Adamic Race. *Genesis* 6:9 tells us that Noah was "perfect in his generations." The Hebrew word for "generations" is "to-lad-aw", which means "ancestry." The correct translation should read, Noah was "perfect in his ancestry," a pure bred without corruption or mongrelization. The other unrelated children of this planet were of the Hu'man Race. The term "hu" refers to the animal or creature, and "man," the potential to become divine.[26] Noah was given instruction from God to keep his lineage "pure" so that the redeemer could be born of the pure Adamic lineage as promised to the upper Adam. Noah, in turn, instructed his children and fore bade outside marriages for all subsequent generations until Abraham's lineage was identified as the lineage of the redeemer. Abraham's lineage had to remain "pure." Apparently this decree remained in effect until the birth of the redeemer.

Modern day Jews have always insisted that they were of a different race who came from another place, a divinely appointed race, and that their DNA is different. The problem is finding a lineage that still remains pure without outside marriages. A DNA study was conducted on a group of Levite males who could trace their lineage back to the Levite priest Aaron, without outside marriages, and some did have a "marker" present not found in other ethnic populations. They are correct in stating that they are a different race, from another place, and were divinely appointed in the beginning, however, there are few bloodlines, if any, that have remained "pure," therefore today we are all considered "one blood." The terms, Race of Adam and the Human Race, are now used interchangeably, denoting the gradual merging into one race.

Wherever the Jews go they are persecuted. Some believe it is because they crucified the Son of God in spite of the world being mostly non Christian. Others say it is because wherever they live they end up controlling most of the wealth. It is generally accepted that they are a stiff- necked people and can be unreasonable at times to deal with. However, the main reason for their misfortune is because of their bloodlines remaining probably one of the most "pure" of all the races. Even though outside marriages have occurred they still remain as a whole a people apart from the general population. This branch of Noah's children has retained some of their divine portion being descendants of Noah who came from another world. They, by remaining "pure" in bloodlines, still possess certain qualities from the spiritual creation. Satan, recognizing these qualities, has extreme hatred toward them. It is Satan who stirs up nations, leaders, and people against them for he desires to exterminate them off the face of this planet in revenge.

The Bible has some mistranslation that has created error in the repopulating of the earth by Noah's children. For example, *Genesis* 10:1-5, 10:20, and 10:31, using the standard King James Version, would have us to believe that everyone is a descendant of one of Noah's children.

Ferrar Fenton's *The Holy Bible in Modern English* translation is translated correctly. In *Genesis* 10:1-5, we read of the descendants of Noah's son Japheth, with the corrected translation as follows:

"From these they spread themselves over the sea coasts of the countries of the nations, each with their language amongst the gentile tribes."

Genesis 10:20, tells of the descendants of Noah's son Ham, correctly translated states:

"These were the sons of Ham, in their tribes and languages, in the regions of the heathen."

Genesis 10:31, correctly translated completes it:

"These are the sons of Shem, by their tribes and by their languages, in their countries among the heathen. The above were the families of the sons of Noah, and their descendants by the tribes. From them they spread themselves amongst the nations on the Earth after the flood."[27]

The Egyptian Gnostics also recognized that other peoples existed before Noah's race as written in *The Apocalypse of Adam (v, 5)* as follows: "Then the seed of Ham and Japheth will form twelve kingdoms, and their seed (also) will enter into the kingdom of **another** people." (bold type added)

From the above translation it now appears that the confusion of languages that occurred at the Tower of Babel was a local occurrence that punished primarily the builders and their families because different peoples had already scattered to various parts of the earth prior to the construction of the tower.

The following was found among some notes one of your authors made, source undocumented, but the notation really seems to fit in this spot, noting what another unknown writer had already observed in the earth's population as follows:

"The tribes which sprang from Noah were **preceded** in their earliest settlements by other tribes whose origin is unknown to us: the Dravitic tribes preceded the Aryans in India; the Proto-Medians preceded the Medians; the Akkadians preceded the Cushites and Semites in Chaldea; the Cannanites were preceded in Palestine by other races. The Black race, just as we find it today, so even in remote times ago, was wholly different from the Caucasian race.

Again the languages of the races springing from Noah are said to be in a state of development different from that in which we find the languages of the peoples of unknown origin."

Shem was entrusted with the bones of Adam and Eve, to be reburied on the temporal earth. *The Zohar* states they were buried in the cave of Machpelah also known as the cave of the patriarchs. Here lie buried the remains of Adam and Eve, Abraham and Sarah, Isaac and Rebecca, and Jacob and Leah.[28] Because the remains of Adam and Eve were buried in this cave, may have determined why Abraham paid such a high price for ownership of the property when his wife, Sarah, died.

Supposedly, when Adam was first buried on the spiritual creation, gold, frankincense, and myrrh were buried with him. These items were also taken aboard the ark and placed with the bones of Adam in the cave of Machpelah. According to the secret book of the Egyptian Gnostics, *The Book of the Cave of Treasures,* Adam is revealed as the first of a long series of prophets, who predicts how the Magi will await the announcement of the Savior, near this cave in which Adam himself will have been interred, and where the Treasures (gold, frankincense, myrrh) are concealed which the Magi will carry to Bethlehem.

Noah gave copies of all the manuscripts or books passed down from Adam, to each of his children. According to *The Book of Jasher*, Abraham studied under the tutoring of Shem for 30 years and learned all his wisdom from these books. He may have taken copies to Egypt for placement in the Hall of Records for preservation and for learning in their universities. It is written in *The Book of Jasher* that wise men and kings sought his counseling in matters of astronomy and astrology. Noah, of course, carried the Rod of God with him all the days of his life and passed it down to Shem. At the foot of Ararat is a city called Naxuana or Nakhichvan, which claims the tomb of Noah. The name means, "here Noah settled."

As written previously, *The Zohar* states that Noah, his family, and many of their descendants were given longevity for generations in order that knowledge of all things, including knowledge of the one true God, could be preserved and taught to the children of the temporal earth. The Elder Gods did also re establish knowledge but much of the knowledge was again lost in various parts of the world due to different circumstances, such as war, famine, and natural disasters, especially in remote regions.

Noah's descendants were given a divine order to go out into the world of the heathen and share their knowledge in all things. Several of Noah's grandsons were given the divine command to re-map the entire world as the cataclysm had changed the world's topography. It took years for them to accomplish the task as they first had to build a large ship. Better ship building was taught to the Hu'man race and the new maps created were to be shared, after all, Noah and his sons were masters of ship building. Soon, the newly drawn world maps, along with knowledge of compass use, enabled mankind to once more take to the open seas. The Phoenicians became master sailors as did the tribe of Dan and an unidentified group called the "sea people." Trade routes were established and unknown parts of the world were explored, colonized, and conquered. Certain sea routes to gold, silver, and copper mines, were tightly guarded secrets. It is known that Egypt had gold mines off the coast of Africa and South America. Australian boomerangs and a stone Olmec head from the La Venta region of Mexico were among the items found in King Tut's tomb whose burial took place in the spring of 1323 B.C. Some of Hapgood's discovered ancient maps may have been copies of the original maps made by Noah's grandsons. It is believed that copies of these maps or charters were studied and used by Columbus to find the New World, however, he really was a late comer on the scene but history credits him for discovering A/merica. It is interesting to note that Columbus never claimed to be the first to cross the Atlantic ocean to discover new lands.

Another grandson of Noah, Votan, was given the divine command to restore knowledge and civilization to the world. In his *Historia del Cielo,* Ordonex de Aquilar says that Votan "proceeded to America by divine command, his mission being to lay the foundations of civilization in that land." Two groups of men accompanied him on his mission. One group was described, as 'faithful soldiers' and the other were the 'shining ones.'[29] The ancient Toltec/Nephite/*Book of Mormon* record identifies who the 'shining ones' were as follows: "And it came to pass that he (Lehi) saw one descending out of the midst of heaven, and he beheld that his luster was above that of the sun at noon-day. And he also saw twelve others following him, and their brightness did exceed that of the stars in the firmament. And they came down and **went forth upon the face of the earth;"**[30] (Bold type added)

From this ancient text it appears that God the son and twelve of his heavenly angels (disciples or teachers) accompanied Votan and others identified as 'faithful soldiers', no doubt unnamed righteous grandchildren of Noah, on the divine mission. Upon the face of the earth they traveled for many years, perhaps even centuries, living among various peoples for unknown lengths of time. They were able to do this because of their long life spans. Myths from the Andes tell us that this group of people came from the south and settled among the people of Lake Titicaca some time after the flood. (Flood of Deucalion) God the son was known to the Aymara-Quichua race of Peru as Viracocha, the creator god, the Maker of All Things, who brought agriculture, animal husbandry, medicine, metallurgy, writing and knowledge of the one true god to the Andean people. Viracocha was described as being tall in statue, of pale complexion, bearded, dressed in a long white robe and wore golden sandals. He claimed divine origin, performed miracles, cured the sick by the laying on of hands, and possessed the power to revive the dead.

To the Mayans, he was known as Quetzalcoatl, Itzamma, Kukulcan, and Gucumatz, the founder of their culture and the first priest of their religion who gave them writing, the calendar, books, and architecture. The writer of *The Codex Vaticanus A* says: "They say that it was he [Quetzalcoatl] who effected the reformation of the world by penance, since, according to his account, *his father had created the world* and men had given themselves up to vice, on which account it had frequently been destroyed. Citinatonali (The Father, The Old One, the Old Serpent covered with green and blue feathers, Creator of All Things) *sent his son into the world to reform it.*" Quetzalcoatl was also known as the *Son of the Virgin,* Coatlicue. Whatever name he was known under, Quetzalcoatl was revered not only as a god, but as a man. The historian Lizana says of him: "He was a king, priest, legislator and ruler of benevolent character."

On most occasions God the son, the 'shining ones', Votan, and the 'faithful' soldiers arrived from the East on ships, other times they entered *by the part of the west.* Legends have confused Votan as being Quetzalcoatl. Votan made frequent visits back to his homeland called by the Toltecs, Tollan-Tlapallan, a country near the eastern ocean. It is said that upon one of Votan's visits to his homeland that he came to a great tower or wall that reached into the heavens, the same tower where the confusion of tongues occurred. As we know, this tower was built in a plain in the land of Shinar, the area where Noah, his grandfather, and his family first settled. Francis Nunez de la Vega, Bishop of Chiapas, states that he saved a partially burned book purported to be written by Votan where Votan wrote "he came to a tower, which had been intended to reach the heavens, a project which had been brought to naught by the linguistic confusion of those who conceived it." Nunez de la Vega states Votan further wrote "that he saw the great wall, namely the Tower of Babel, which was built from earth to heaven at the bidding of his grandfather Noah." Father Ordonez Aquilar, deriving his knowledge from the same partially burned book, states that, "Votan built a temple by the Huehuetan river, which was called the "House of Darkness", wherein he deposited the national records."[31] (Yucatan hall or records)

The Tahitians remember three great ships that sailed to an out-lying island with oarsmen, whose paddles looked like centipede's legs and giant sails like enormous birds. To their amazement a Fair God, with gray-green eyes and long reddish curly brown hair, in a white robe approached them walking on water. The Fair God they called Wakea, the Healer. He was known as a prophet and teacher who came among the Polynesian Islanders sometime in the first century of the Christian era.

The Pawnee remember him as Paruxti, a prophet who came and taught them of His Father, Tirawa. To the Dakotah he was known as the Great Wakona who taught them the rite of baptism and foretold the coming of white men. Among the Choctah, he was known as Ee-me-shee. To the Chinooks his name was Tla-acomah, meaning Lord Miracle Worker. He was known as Tla-acoma to the Pueblos and as the Great Ta-copah to the Wallapai Tribe. He asked every tribe that he visited and taught to name him in their own tongue. The Pale God traveled over much of Mexico, North America into Canada, South America, and Polynesia on this side of the globe. It appears that on some occasions the Pale God traveled alone.[32]

On the other side of the globe, stories have surfaced where the Shining Ones appeared in the Asian Mountains, restoring knowledge in China and Tibet. According to the *Encyclopedia Americana* "Fuh-hi, who is regarded as a demigod, founded the Chinese Empire 2,852 B.C. He introduced cattle, taught the people how to raise them, and taught the art of writing."

The white god, Quetzalcoatl, promised he would someday return to Mexico, and he indeed kept his promise. His visit was recorded in the *Book of Mormon, III Nephi* chapters 11 through 17, wherein at least two thousand five hundred witnesses gave their collective testimonies. The visit occurred the 6th day of Passover, 6 Ahau 3 Mak, Sunday, March 26, A.D. 34.

The ancient Toltec/Nephite record was carried into North America by the last Toltec/Nephite warrior prophet named Moroni in about 385 to 390 A.D. He buried the record in a stone box under a hill known as Cumorah, which is the ancient name of the first known record keeper in the New World. His name was shortened to the brother of Jared. His true name was Mhanri Moriancumr. Centuries later, the angel Moroni led Joseph Smith, Jr. to their hidden location and recovery. Joseph then translated that record directly into English by the gift and power of God. It has been translated into more languages and distributed to more countries than any other book except *The Bible* and is known around the world as the *Book of Mormon.*

When Moctezuma encountered Cortez for the first time in 1519, he mistook him for the bearded white god, Quetzalcoatl, returning to Mexico. Because the ancient Toltec/Nephite record had been carried far into North America centuries earlier, the early inhabitants and their descendants of that region had lost all knowledge of Quetzalcoatl's recorded visit. Only legends remained. They are still looking for the return of the bearded white god, but he has already been there and left again.[33]

Genesis 10:25 and 1 *Chronicles* 1:19 both state the following: "And unto Eber were born two sons: the name of the one was Peleg; because in his days the earth was divided:" (Peleg was the fourth great grandson of Noah.) According to Strong's Concordance "Peleg", in the Hebrew language, means a dividing by a "small channel of water" and also its root is associated with the meaning of an earthquake. *Genesis* 11:10-17, says that Peleg was born 101 years after the flood of Noah (Deucalion Flood).

The division of the earth was not an act of division by the inhabitants of the earth by tribes and peoples, as this had already occurred, or a breaking asunder of the continents. It would be absurd to think that all the continents divided at once. The verses do say the earth but not the entire earth. To do so would cause another major cataclysm that would nearly destroy the earth's population again. Fenton's Modern English Bible states, "...because in his (Peleg's) days the continent was split up."[34] Note the word *continent,* not *continents.* The following is what we theorize is the most likely explanation of those verses. In 1930-31, a Babylonian clay tablet was found in the excavated ruined city of Ga-Sur or Yorghan Tepe, Mesopotamia, near the towns of Harran and Kirkuk, 200 miles north of the site of Babylon or present day Iraq. It is calculated that the map was drawn 2,300-2,500 B.C. Later another date was given as 3,800 B.C. On the clay tablet, which would fit in the palm of your hand, is a drawn map called the Nuzi Map.

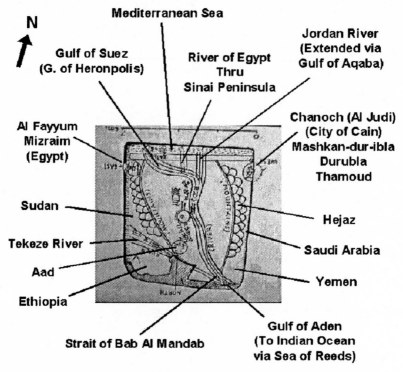

N

Mediterranean Sea

Gulf of Suez
(G. of Heronpolis)

River of Egypt
Thru
Sinai Peninsula

Jordan River
(Extended via
Gulf of Aqaba)

Al Fayyum
Mizraim
(Egypt)

Chanoch (Al Judi)
(City of Cain)
Mashkan-dur-ibla
Durubla
Thamoud

Sudan

Hejaz

Tekeze River

Saudi Arabia

Aad

Yemen

Ethiopia

Strait of Bab Al Mandab

Gulf of Aden
(To Indian Ocean
via Sea of Reeds)

Redrawing of Ancient Nuzi Map from: http://www.henry-davis.com/MAPS/AncientWebPages/100D.
html as viewed from Brian's Annex website.

This map shows the Edenic Valley before it was flooded by the Red Sea. The map shows the River of Egypt flowing through the Sinai Peninsula from the Mediterranean Sea and the Jordan River (extended via the Gulf of Aqaba) joining together into one river that flows through the entire valley emptying into the Gulf of Aden. (to Indian Ocean via Sea of Reeds) Today that valley is the Red Sea. We believe that at one time Egypt and Arabia were joined by this valley landmass that sunk in the days of Peleg due to either an earthquake or tectonic plate movement. Earthquakes are common in the Mediterranean area, the dates would correspond to the days of Peleg, and it was in the general area where Peleg lived. However on the other hand, geologists contend that the great rift occurred approximately 20 million years ago, the result of earthquake activity. The Red Sea is really a great crack in solid rock that has filled with water. Its main body length is approximately 1, 200 miles long and about 200 miles wide. Shown on the map is a southern city in the submerged part of the valley that may be the city (tribe) of Aad as a compatriot city of Thamoud, located in the upper northeast, described throughout the Qur'an, whose people led sinful lives and that both were destroyed together.[35] When you look at this area on a map it appears that this stretch of land just pulled away leaving the Sinai Peninsula barely connecting Egypt and Saudi Arabia. In fact, if you closely examine a relief map of the area, you can see where a mountain range was sheared and dragged apart. In time the Peninsula may completely separate. Geologists call this area the Afar Triangle. The Afar fracture is a three way split. The area on the African continent running south from that fracture locus is the Great Rift Valley. The area going northward runs along the bottom of the Red Sea and up into the Dead Sea area of Israel.

Many have tried to locate the Garden of Eden on this temporal earth but it never existed on this planet. It was located on the spirit earth. When the two planets merged together, the approximate former location of the Garden on the spirit earth would now be re- positioned on the temporal earth, at a place called Spring Hill, Daviess County, Missouri. (*D & C.* 116)

Around 2,000 B.C., the Greek culture began to rise. Ancient Greeks loved beauty in all things that was reflected in their architecture, statues, poems, theater, politics and government. Most of their magnificent temples were shrines built to honor the Elder Gods and Titans. Were they hazy reminiscences of past occupations or did they continue to have physical contact with these gods? We probably will never know for sure but one thing occurred for certain, some of the children of men continued to have giant offspring for generations. One of the most famous and beautiful temples built during this period of time was probably built by the Greeks in the megalithic city of Ba'alBek, now Lebanon. On the ruins of an earlier megalithic building foundation, they constructed a magnificent huge temple honoring the god Jupiter. Ancient Greece had temples dotting the countryside and coast lines everywhere. It was a religion that spread into surrounding countries and remained until Christianity eventually replaced, and in some cases merged with the old beliefs. Even Rome is said to have been founded in 795 B.C., by Romulus and Remus; twin sons of Mars, the god of war and Rhea Silvia, a mortal vestal virgin. She was sentenced to be buried alive for breaking her vows. Be that as it may. **The Rest is History!**

Allan & Delair state our concluding remarks to this chapter best from *When The Earth Nearly Died,* "It is thus scarcely surprising that Earth is far from being the most perfectly functioning planet that a surprisingly large number of people assume it to be. On the evidence of its fluctuating motions and behavior patterns, it has not merely recently sustained serious disruption to many of its natural physical and atmospheric features but is still slowly recovering from the traumatic event which imparted them."[36]

End Notes

[1] *Cataclysm!: Compelling Evidence of a Cosmic Catastrophe in 9500 B.C. by D.S. Allan & J.B. Delair, Bear & Co., a division of Inner Traditions International, Rochester, VT 05767 Copyright © 1995 & 1997 by D.S. Allan & J.B. Delair. (pp. 14-15)*

[2] *The Atlantis Blueprint*, ©2000 by Rand Flem-Ath and Colin Wilson. Published by Delacorte Press a division of Random House, Inc. , NY, NY. p. 209)

[3] Excerpts taken from *The Return of The Serpents of Wisdom* ©1997 by Mark Amaru Pinkham. Published by Adventures Unlimited Press, Kempton, ILL. All rights reserved. Used by permission of the publisher, pp.30, 31, 34.)

[4] (*Return of Serpents of Wisdom*, p. 93, as cited in *The Mysteries of Ancient South America*, Harold Wilkens, 1946, Citadel Press, NYC)

[5] (Ref; *The Gold of the Gods*, Erich von Daniken, Bantam Books NY, 1972, pp.2,3,68,9,10.)

[6] (*Serpents of Wisdom*, pp.54, 95)

[7] Greek story of *Philemon and Baucis* (p. 328)

[8] (eglossary@thesociety.org viewed Dec. 5, 2002)

[9] *Atlantis Mother of Empires,* ©1939 by Robert B. Stacy-Judd. Published by Adventures Unlimited Press, Kempton, ILL. (p. 126)

[10] (Eusebius, P. 1913, *Werke* (Leipzig), tr by R. Helm, 9 vols; Vol. v, "Die Chronik"; Frazier, J G. 1918, *Folklore in the Old Testament*, London, 3 vols; Vol. 1, p. 159)

[11] (Way, A. 1924, *The Riddle of the Earth,* London, 251 pp; chap xi, pp160-161)

[12] *Cataclysm!: Compelling Evidence of a Cosmic Catastrophe in 9500 B.C.* by D.S. Allan & J.B. Delair, Bear & Co., a division of Inner Traditions International, Rochester, VT 05767 Copyright © 1995 & 1997 by D.S. Allan & J.B. Delair. (p. 263)

[13] (*Cataclysm!: Compelling Evidence of a Cosmic Catastrophe in 9500 B.C.* by D.S. Allan & J.B. Delair, Bear & Co., a division of Inner Traditions International, Rochester, VT 05767 Copyright © 1995 & 1997 by D.S. Allan & J.B. Delair., p. 256, as cited from Frobenius, L. 1898, *Die Weltanschauung der Naturvolker* (Weimar), xv + 427 pp; see pp355-7.)

[14] (*Book of Job*, chap 38, v 29-32)

[15] (*When the Earth Nearly Died*, et al p. 101)

[16] (*Gods From Outer Space,* Erich Von Daniken, Bantam Books, NYC, 1970, p. 147)

[17] et al (pp.161,161)

[18] (*Gods From Outer Space,* Erich Von Daniken, Bantam Books, NYC, 1970, p. 148.)

[19] (et al p. 1006-1007)

[20] (et al p. 68-69) Parenthesis added.

[21] *Atlantis Mother of Empires,* ©1939 by Robert B. Stacy-Judd. Published by Adventures Unlimited Press, Kempton, ILL. p. 85)

[22] (*Zohar* Vol. 11, p. 104)

[23] Excerpted from *Sepher Rezial Hemelach: The Book of the Angel Rezial*, edited and translated by Steve Savedow with permission of Red Wheel/Weiser, York Beach, ME and Boston, MA. To order call 1-800-423-7087, pp.79, 213- 215) Excerpted from *The Hebrew Book of the Dead In the Wilderness* ©2003 by Zhenya Sunyak. Published by Tikin Press, Hallandale, FLA. Used by permission of the publisher. All rights reserved. pp.8,9,10, 208, 209)

[24] *Book of Enoch*, chapter 8.

[25] KJV *Hebrews* chapter 5.

[26] (*The Christ Consciousness*, Norman Paulsen, The Builders Publishing Company, SLC, 1984 p. 294)

[27] *The Holy Bible in Modern English,* translated by Ferrar Fenton. Published by Destiny Publishers, P O Box 177, Merrimac, MA 01860-0177.

[28] (*The Zohar,* translated by Simon, Sperling and Levertoff, ©1984, London, The Soncino Press, Ltd., *Vol. 11,* translated by Harry Sperling and Maurice Simon, The Soncino Press, NYC-London, 1984, pp13, 386)

[29] (*The Atlantis Blueprint*, ©2000 by Rand Flem-Ath and Colin Wilson. Published by Delacorte Press a division of Random House, Inc., NY, NY. as cited in Flem-Ath, Rand and Rose, *When the Sky Fell*, p. 28)

[30] (*1 Nephi* 1:9-11, bold type added)

[31] (*Atlantis-Mother of Empires,* et al p. 98)

[32] (*He Walked The Americas,* L. Taylor Hansen, Amherst, Wisconsin, 1963)

[33] *New Evidences of Christ in Ancient America*, Waren & Brown, Book of Mormon Research Foundation, P. 169

[34] *The Holy Bible in Modern English,* translated by Ferrar Fenton. Published by Destiny Publishers, P O Box 177, Merrimac, MA 01860-0177.

[35] (*Qur'a*n Sura 69:1-10; Sura 89:6)

[36] *Cataclysm!: Compelling Evidence of a Cosmic Catastrophe in 9500 B.C.* by D.S. Allan & J.B. Delair, Bear & Co., a division of Inner Traditions International, Rochester, VT 05767 Copyright © 1995 & 1997 by D.S. Allan & J.B. Delair. (p. 190)

The Future World

Mark 10:30

"But he shall receive an hundredfold now in this time, houses, and brethren, and sisters, and mothers, and children, and lands, with persecutions; and in **the world to come** eternal life."

Luke 18:30

"Who shall not receive manifold more in this present time, and in **the world to come life** everlasting."

Hebrews 2:5

"For unto the angels hath he not put in subjection **the world to come**, whereof we speak."

Hebrews 6:5

"And have tasted the good word of God, and the powers of **the world to come…. .**"

D&C 59:23

"But learn that he who doeth the works of righteousness shall receive his reward, even peace in this world and eternal life in **the world to come**."

D&C section 63:48

"He that sendeth up treasures unto the land of Zion shall receive an inheritance in this world, and his works shall follow him, and also a reward in **the world to come**."

Qur'an The Cave 18:107

"As to those who believe and work righteous deeds, they have, for their entertainment, the **Gardens of Paradise**."

Qur'an The Believers 23:11

"Who will inherit **Paradise**; they will dwell therein {forever}." (note emphasis added to all above scriptures)

The day when the earth is returned to its former state of "Paradisiacal Glory" has been eagerly anticipated down through the ages. It is described as **the world to come or Paradise**, where death and pain is unknown and its righteous residents live forever young in a state of eternal bliss as testified in the following biblical scripture:

"And God shall wipe away all tears from their eyes; and there shall be no more death, neither sorrow, nor crying, neither shall there be any more pain: for the former things are passed away."[1]

It was not until *New Testament* times that the belief of a new world to come was taught and embraced first by Christians, then Muslims, as confirmed in the above opening scriptures. Although the *Old Testament* doesn't mention the concept of a *new world to come* and only foreshadowings of a resurrection, the concept of a resurrection and a world to come was taught in ancient Jewish Kabbalah. There are many references in *The Zohar* that attest to the belief. For example in Vol. 111, p. 75, in *The Zohar* is the following writing:

"the Lord designed the worlds, this world and the world to come."

In the same volume, p. 93, is found the following *Old Testament* verse with interpretation from Rabbi Hiya:

"Thy dead ones will live,"[2] It is evident that not only will there be a new creation, but that the very bodies which were dead will rise on the Resurrection Day.

The last example found in the same book, p. 193, is found the following writing:

"Blessed are the righteous who meditate on the *Torah* day and night! Blessed are they in this world, and blessed are they in the world to come!".

However, before the event of the world to come can take place, it is a common belief that the earth must undergo a destruction, death and a resurrection. When these phases are completed a new world emerges of indescribable beauty, a paradise returned for the righteous to inhabit forever. The following scriptures document these beliefs:

"But the day of the Lord will come as a thief in the night; in the which the heavens shall pass away with a great noise, and the elements shall melt with fervent heat, the earth also and the works that are therein shall be burned up."[3]

"Behold, will you believe in the day of your visitation— behold, when the Lord shall come, yea, even that great day when the earth shall be rolled together as a scroll, and the elements shall melt with fervent heat, yea, in that great day when ye shall be brought to stand before the Lamb of God—then will ye say that there is no God?"[4]

"And every corruptible thing, both of man, or of the beast of the field, or of the fowls of the heavens, or of the fish of the sea, that dwells upon all the face of the earth, shall be consumed;

And also that of element shall melt with fervent heat; and all things shall become new, that my knowledge and glory may dwell upon all the earth."[5]

"Then contemplate (O man!) the memorials of Allah's Mercy!—how He gives life to the earth after its death: verily the same will give life to the men who are dead: for He has power over all things."[6]

"It is He who brings out the living from the dead, and brings out the dead from the living, and who gives life to the earth after it is dead: and thus shall ye be brought out (from the dead)."[7]

"It is Allah Who sends forth the winds, so that they raise up the clouds, and we drive them to a land that is dead, and revive the earth therewith after its death: even so (will be) the resurrection."[8]

The Hindus believe that the destructive phase includes the entire universe and refer to this final consumptive phase as "flames belched forth from the fangs of Shesha" which incinerate all material forms in the universe.

Mark Pinkham states in his book, *The Return of The Serpents of Wisdom*, the following:

"During the Destruction Phase, the Primal Serpent in the form of a Fire Serpent completely annihilates the universe and reverts all material form back to pure Life Force or Cosmic Fire.

The final destruction of the universe culminates in the ascent of the Serpent up the cosmic tree and its re-emersion in the cosmic sea. Then during the ensuing "in-between" or Pralaya Phase, the Serpent (now in a dormant/potential state) will drift peacefully upon or with the cosmic sea while awaiting the re-creation of the universe.

The ancient Egyptians of Heliopolis alluded to the universal dissolution as "the end of time (when) the world will revert to the primary state of chaos and Atum (the Creator Dragon) will once again become a Serpent (pure Life Force).

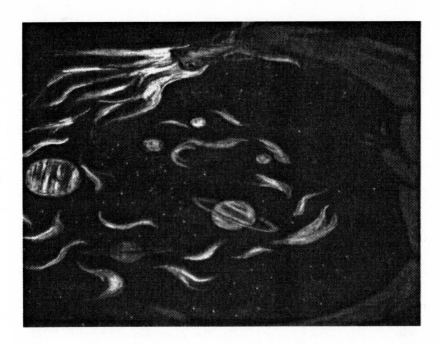

The Primal Serpent was the first tangible form assumed by Spirit, and was the vehicle of all God's powers, including the triune powers of creation, preservation and destruction. Through it, God created the entire universe."

Will our Universe eventually self destruct? Readers are referred back to the chapter, *The Eternal Heavens*, where the oscillation theory was discussed. In a nutshell, Buddhism, Janinism, Hinduism, and Brahmanism view that our present universe is only one in a beginningless series, reappearing and disappearing from time to time. Remember the word temporal means temporary; that the universe had a beginning and will some time end.

Whether or not the entire universe is destroyed or just the solar system that contains the earth is not made clear by biblical scriptures or other sacred writings, however the promise of a new heaven and new earth is given in the following scriptures:

"Looking for and hasting unto the coming of the day of God, wherein the heavens being on fire shall be dissolved and the elements shall melt with fervent heat?

Nevertheless we, according to his promise, look for new heavens and a new earth, wherein dwelleth righteousness."[9]

"And I saw a new heaven and a new earth: for the first heaven and the first earth were passed away; and there was no more sea."[10]

"One day the earth will be changed to a different earth, and so will be the heavens, and (men) will be marshalled forth, before Allah, the One, the Irresistible."[11]

Just how will the present earth be changed to a different earth and different heavens? Readers are referred back to the chapter, *The Arrival of Noah*, for when Brigham Young in a sermon stated:

"When this world was first made it was in close proximity to God. When man sinned it was hurled million of miles away from its first position, and that was why it is called the Fall."

And:

"This earthly ball, this little opake {opaque} substance thrown off into space, is only a speck in the great Universe: and when it is celestialized it will go back into the presence of God, where it was first framed."[12]

Joseph Smith taught that when the earth is glorified, it "will be rolled back into the presence of God and be crowned with celestial glory."

John Taylor, the third Mormon prophet, was more specific. He taught that our earth has "fled and fallen, from where it was first organized, *near the planet Kolob.*" (note: Kolob is a star)[13]

The only way that the earth can return back to its origin, in a high spiritual heaven, near to the throne of God and near the giant star Kolob, is for its spirit (the spiritual or airy earth) to leave or depart the dead, dying, or destroyed temporal earth. It will "roll back" the same way it came passing through the mighty gates of heaven and rapidly passing through millions of years of time and dimensions, probably traveling through the fourth dimension. When it reaches its final destination, it will be glorified or celestialized. The heavens surrounding the spiritual creation will be different from its temporal abode and appear as new. *The Zohar* speaks of the light of the sun and moon returning to their former brightness. *Isaiah* 30:26 also speaks of the same event in the following scripture:

"Moreover the light of the moon shall be as the light of the sun, and the light of the sun shall be sevenfold, as the light of seven days, in the day that the Lord bindeth up the breach of his people, and healeth the stroke of their wound."

The spirit earth will be as it was when first created before it fell, only now glorified.

The highest heaven or universe is located in the third heaven, seen by the apostle Paul in vision, that is further divided into three heavens or universes. Readers are referred back to the chapter, *The Heavens*, for more on the divisions of the celestial heaven.[14]

Just before the crucifixion of Jesus he said the following to his apostles:

"In my Father's house are many mansions; if it were not so, I would have told you. I go to prepare a place for you.

And if I go and prepare a place for you, I will come again, and receive you unto myself; that where I am, ye may be also."[15]

It appears that the Savior went to prepare a place in a high spiritual heaven, near to the throne of God, among other glorified stellar and planetary worlds for the eventual return of the glorified spiritual earth and its celestial inhabitants. There it regains its former title as one of the many footstools of the Lord.[16] Adam and his righteous descendants will be given back dominion of the celestial earth, as promised by God. The physical or temporal earth is destroyed, leaving Satan and the remaining "fallen angels" awaiting judgment for their sins and crimes against mankind on this planet.

Some time during the next eighteen centuries, the resurrected Savior accomplished the promised task when he revealed the following latter day revelation in 1833:

"I *have* prepared a place for you; and where my Father and I am, there ye shall be also."[17]

When will the predicted end of the earth or end times take place? It is the general consensus of our times that we are now living in the last of the last days. The following are some ancient prophecies that speak of these days.

Prophecy of the Cherokee Rattlesnake Constellation

Upon the Heavens design, within the Cherokee Rattlesnake Constellation, it is written. And upon the Rings of *Time Untime* of the Cherokee Calendar it also says **Behold Rattlesnake Constellation**, and remember.

The Rings of the Cherokee Calendar are read by the winding motion of the Rattlesnake. To add or to subtract *Time Untime,*, to move the wheels backward or forward, to add a wheel to move the rings, to tell the tale what was, what is, and to be, the Rattlesnake and its design tells the movement of the *Rings of Time.*

And in the sky heavens the motions of the movement of stars, planets, and life tell the tale of its hearing. For all is as the winding of the Rattlesnake. And in the year 2004 the Morning Star shall be first and in the year 2012, the Evening Star shall be first. And upon those years the crown of the Feathered Serpent shall bear its colors and honor. The hands shall hold the bowl and the tail shall be as the roots of a tree, The Pleaides Tree of the beginning.

And in the year 2004 and 2012, shall be the *Time Untime* of the Feathered Serpent of the Sky Heavens. And the Rings shall turn upon those years of prophecy foretold on the Rings of *Time Untime.* And Ywahoo Falls Kentucky shall sing of Venus and the Feathered Rattlesnake.

For this is not all the prophecy and not all the things thereof. For if it was meant for one to know and I give only part, then you have received a great gift. For if I tell all and it was not meant for one to know that I have wasted my breath. For this is a sacred thing, the Venus Alignment and the Feathered Serpent of the Heavens.

And in the year 2012, the Cherokee calendar ends. And all is reborn. For the Feathered Rattlesnake comes and shall be seen in the heavens in the years 2004 to 2012. The Rattlesnake Constellation will take on a different configuration. The Snake itself will remain, however, upon the Rattlesnake shall be added upon its head feathers, its eyes will open and glow, wings spring forth as a winged Rattlesnake, it shall have hands and arms and in its hands shall be found a bowl. The bowl will hold blood. Upon its tail of 7 rattles shall be the glowing and movement of the Pleiades. The Rattlesnake shall become a Feathered Serpent of *Time Untime.*

In the south of the americas…it is related as the coming of Quetzalcoatl. The ancient Cherokee relate it as the coming of the Pale One once again.[18]

The Hopi Prophecy of 7 Thunders

In the early 1980's, there was a meeting of Elders from the Algonquin People of the northeast. They spoke to Elders of various Nations, and all these Nations gathered, brought their prophecies, and talked of the time of change. The meeting, called the "Rainbow Walk," took place in the Taos Mountains.

From this meeting came a kind of compilation of Native American prophecies from many tribes. As things start to break down, the discovery of many things kept hidden until this time start to appear, such as the secrets of the pyramids, ancient texts (rediscovered), etc. The following is the prophecy of the 7 Thunders as related by Robert Ghost Wolf.

"Native Americans still hold the oldest existing on going knowledge. Knowledge passed on for thousands of years. Truth. A message from the heart. The prophecies of the seven thunders are about these times. The first three have already occurred. This is a summary of the prophecies. The fourth is nearly fulfilled.

1. The people of the world will experience mass starvation, sickness and homelessness.

2. The sky will become sick with holes like sores. From these holes bacteria will come into the waters. This will affect our physical and dimensional realities.

3. New species and mutations will occur. Mutations in humans will start to happen.

4. Two comets-the twins in the heavens, one guards the north pole, one the south. They will cause the poles to shift and our realities will shift too. New information will be unearthed, lost cities will be found. The Devil winds will blow and the Earth will rid her toxins. Four great women will perish which will signal the

beginning of seven circles which are the answers to the seven thunders.

5. The atmosphere can no longer repair itself. Time and reality as we perceive it will shift. Good and evil will emerge. Man will experiment with the Earth and cause a volcanic eruption in Mexico and a chain reaction of earthquakes will go up the Cascade Mountain Range.

6. The people from the stars will return to help us. The Earth has picked up the consciousness of the people and it is on self-destruct.

7. The Purifier comes-The Twelfth planet will pass the Earth, causing three and a half days of darkness in the eclipse. The pyramids clock, keeper of the Earth, runs on sun light. After the darkness we will emerge into the fifth world." (note: *Revelation* chapter 10 speaks of the 7 thunders)

Prophecy of the Red Heifer

Tradition records that a red heifer, born in Israel, in our generation is a herald of the Messianic era. Apparently a red heifer was born about March 2001, in Israel. This heifer could be a candidate to be used in the process of purification described in the book of *Numbers*, chapter 19, and is believed to be a prerequisite for the rebuilding of the Holy Temple. This red heifer is believed to be the first born in Israel in at least 2,000 years.

Zulu Prophecy

Let me tell you two last things please. One, it is this, that I am told by the great storytellers of our tribes, that fresh water is not native to our earth. That at one time, many thousands of years ago a terrible star, or the kind called mu-sho-sho-no-no, the star with a very long tail, descended very close upon our skies. It came so close that the earth turned upside down and what had become the sky became down, and what was the heavens became up. The whole world was turned upside down. The sun rose in the south and set in the north. Then came drops of burning black stuff, like molten tar, which burned every living thing on earth that none could escape. After that came a terrible deluge of water accompanied by winds so great that they blew whole mountain tops away and after that came huge chunks of ice bigger than any mountain and the whole world was covered by ice for many generations.

After that the surviving people saw an amazing sight. They saw rivers and streams of water that they could drink and they saw that some of the fishes that escaped from the sea were now living in these rivers. That is the great story of our forefathers. And we are told that this thing is going to happen again very soon because the great star, which is the lava of our sun, is going to return in the day of the year of the Red Bull, which is the year 2012.[19]

Incan Prophecy

Willaru Huayta, a Peruvian spiritual messenger, says that 2013 is the end of the Inca calendar, and in that year, a "huge asteroid", three times larger than Jupiter will pass close to earth, causing cataclysms that will kill off most of humankind.

The Incas refer to the end of time as we know it as to the death of a way of thinking and a way of being, the end of a way of relating to nature and the earth. There will be tumultuous changes happening in the earth. After a period of turmoil and purification, a new human will emerge into a golden age, a golden millennium of peace.

Mayan/Aztec Prophecy

According to the *Dresden Codex,* one of the few surviving documents of Mayan culture and religion, not destroyed by the Spanish conquerors, is The Prophecy of the Six Suns. According to the Codex, the history of the world was divided into six ages of man. Each age, or sun, died through a cycle of destruction and the recreation of a new civilization.

The death of the Fifth Sun occurred on July 11, 1991, during a solar eclipse. In Mexico City, seventeen different camcorders recorded the same UFO, visible for 23 minutes. A nearby volcano, Mount Popocatepetl became active right after the eclipse. These events occurred in accordance to what was written in the Codex, 1,236 years ago. As far as the Mayans are concerned, there is no Seventh Sun. Their calendar and history ends for them on December

21, 2012, when a rare event known as the "precession of the Equinox" will begin.

This rare astronomical alignment is the slow process by which the winter solstice sun comes to conjoin the "dark rift" in the Milky Way, a place called by the Mayans, the womb of the cosmic mother. A full cycle of this phenomenon is completed in about 26,000 years. The date of the alignment is the end date of the 13-Baktun Great Cycle, a cycle of approximately 5,125 years or December 21, 2012, in the Gregorian calendar.

According to the prophecy earth will pass inside the center of a magnetic axis and that it may be darkened with a great cloud for 60-70 hours. The earth may not be strong enough to survive the effects. It will enter another age, and when it does, there will be great earthquakes, floods, volcanic eruptions, great illness, and other serious events. Few survivors will be left. It is the death of the Sixth Sun. The Mayans view it as a rebirth, the start, once again, from zero point of world time, a major transition point, and the creation of a New World Age.

It will be the start of a new era resulting from and signified by the solar meridian crossing the galactic equator, and the earth aligning itself with the center of the galaxy. For the first time in 26,000 years, the sun will rise to conjunct the intersection of the Milky Way and the plane of the ecliptic, forming a cosmic cross within the Milky Way. The Milky Way was viewed as the cosmic Tree of Life or World Tree. When the Serpent or Snake Constellation crosses or conjuncts into the Milky Way, it is viewed as the Serpent

(Primal Dragon or Life Force) climbing up the Cosmic Tree and its re-emersion into the Cosmic Sea. The Cosmic Sea is the "dark rift" in the Milky Way and the location they believed to be the place of creation or "womb." The Vedics, Cherokee, Maya, and ancient Egyptians, call this ascent of the Serpent, Primal Dragon, or Life Force the final destruction of the universe, or destructive phase. In the Garden of Eden, when the Serpent (Primal Dragon or Life Force) descended the cosmic tree, these ancients believed it started the creation of this universe and earth, or creative phase.

The Aztecs adopted the Mayan calendar and modified the calendar somewhat into five ages with different names. They believe that the fifth age and final Sun, "the Age of the Jaguar" comes to an end on December 21, 2012.

Vedic Prophecy

According to Vedic tradition, we are in the fourth world age, or Kali Yuga (Dark or Iron Age). Hindus are awaiting the coming of the Kalki Avatar at the end of this present age. Bahai's believe that the Kali Yuga age has already ended and, as promised in the *Bhagavad Gita*, the Lord has again manifested Himself to humanity as the Baha'u'llah or Kalki Avatar. The coming of the Baha'u'llah is the start of the Sat or krta Yuga, the Golden Age. It will be a time when the people will return to righteousness and the world will be at peace.

The Kali Yuga age is the furthest from the celestial seat of Brahman and is a predicted time when humans will be most wicked, greedy, lustful, violent, disrespectful of parents and laws, unchaste, and unbelievers. It is a time foretold in the Hindu scriptures which the Bahai's state have already been fulfilled in this age. The Bahai's say what is described in the Hindu books is exactly what we are seeing in the world today.

The Bahai scholars believe that the exact date of the end of the Kali Yuga and the coming of the Kalki Avatar is 1844, and is also the year of the beginning of the Bahai Faith. The Hindus believe that the Kali Yuga age ended in 1898, and that the Lord Vasudeva (Vishnu) will become incarnate here in the universe in the form of Kalki Avatar sometime in the future. The Bahai's believe that the Kalki Avatar has already come, and that the Kalki Avatar is the Baha'u'llah.

As foretold in the Hindu scriptures and in the Vishnu Purana, the Kakli Avatar or the Baha'u'llah will establish righteousness and new teachings upon the earth and the remnant of mankind will be transformed into a new type of human, with new ideals and spiritually awake. The Golden Age will be established.

Christian Prophecy

Most Christian faiths believe that in the last days that there will great signs in the heavens. These signs are harbingers of the second coming of Jesus Christ. Prior to his coming there will be the Rapture of the just, then tribulation followed by three days of darkness and an universal shaking of the earth. This shaking will come from above the earth and not beneath. The earth will undergo a purging or cleansing process by fire. Christ will then establish his kingdom on earth for one thousand years with the survivors of the tribulation and three days of darkness.

From the Catholic view point they view the end time events as the time of the Great Chastisement which will be greater in magnitude than the great deluge of Noah.

A cross will appear in the sky. It will be a sign that the final events are near. When the prophesied events begin, the night will be bitterly cold, the wind will howl and roar, then will come lightning, thunderbolts, earthquakes, the stars and heavenly bodies will be disturbed and restless. There will be no light, but total blackness. Hurricanes of fires will rain forth from heaven and spread over all the earth, fear will seize mortals at the sight of these clouds of fire, and great will be their cries of lamentation, many godless will burn in the open fields like withered grass. Seventy-five percent of the earth's population will be lost. During the three days of darkness only blessed candles will be able to give light. To stray outside ones home will be instant death. Window shades are to be pulled, even looking out the window will be instant death. People are advised to pray for the world and for its people.

As an act of mercy, a comet (the Ball of Redemption) will be sent to earth to bring it back to the old alignment with the sun and the moon. The continents will be rearranged. The Lord will then establish his kingdom on earth for the millennial reign for one thousand years, a reign of peace.

It is not the intent of the authors to preach *Doomsday Prophecy,* but rather point out the thread of similarity in all these prophecies that indicate something big is going to happen in December 2012.

Mayan calander showing the world's end in 2012

We have heard little about this upcoming event in the news media but certainly in a few years there will be more on CNNews. Doomsday prophets will be everywhere but what do today's astronomers have to say about this event? Quiet frankly, very little at present, however there are some interesting web sites that explain the "Precession of the Equinox" in detail along with Mayan beliefs. It is quiet amazing that the ancient Mayans, Olmecs, and Egyptians understood this event and that the Mayans were able to develop a calendar that could accurately pin point this event down to the day and year several thousand years in the future. On the lid of the sarcophagus of the giant Mayan king, Pacal, is engraved the stylized World Tree in the form of a cross. Pacal is shown either descending or ascending the tree, depending on how one interprets the famous carving. It seems that even in death Pacal had carved in stone a record fore telling the event, and perhaps it even foretells the time of his resurrection associated with the alignment and event of 2012.

The sarcophagus lid of Pacal

The astronomers have just recently confirmed the existence of a black hole at the galactic centre or "dark rift" in the Milky Way, something the ancient Mayans knew as they even had a glyph to represent the black hole at the galactic centre.

This black hole has been described as "a potent energy portal, an unobstructed flow of energy and a very powerful vortex or stargate." The question is…does this black hole generate enough energy to destroy our sun and earth once aligned with it or can it simply cause disturbances of cataclysmic proportions? Another possible scenario is when the sun and earth align with the opening or "vagina" in the Milky Way leading to the black hole or "womb", can the black hole cause disturbances to the sun's surface or corona that will cause the sun's rays to flare millions of miles further and hotter into space igniting the earth with fire? SOHO satellite monitoring of the sun has accurately mapped more than 530 sunspots, some of which when facing the earth emit X-class cosmic radiation. The aurora borealis is a visual indicator of the reaction between these charged particles and the upper atmosphere. There are about 50 to 70 sunspots facing the earth each year, except in 1957, when more than 190 occurred. On January 28th, 2004, there were no sunspots facing earth. This is extremely rare. These are strong magnetic convergences that result in cool spots in the surface of the sun that can release huge solar flares millions of miles into space if they become unstable. When this occurs, there can be interruptions in radio communications from geo-magnetic storms, heavy showers of cosmic rays, and even enough radiation to sear the surface of the earth. An X8 solar flare, which occurred from sunspot number 76 on November 3rd of 2003, accelerated the solar wind to more than 1,000 kilometers per second and showered the earth with powerful cosmic radiation. The proton fluence greater than 1MeV was 8.6×10^8, greater than 10MeV was 8.9×10^7, and greater than 100MeV was 2.2×10^5 protons per square centimeter per day.[20] That means billions of highly charged protons poured into every square centimeter of the atmosphere from the sun. When X-class cosmic radiation occurs, passenger jets reroute away from the Arctic Circle to avoid excess radiation. This prospect is spine-chilling and more real than the public knows.

As stated in *Fingerprints of the Gods,* " Scientists surmise that the solar system has a certain cosmic-electrical balance mechanism or force that extends a billion miles from the centre of our solar system. Such an electrical balance or force is not accounted for in current astrophysical theories."[21]

There is a group of sunspots known as the monster group. They are large and lately have offered surprising behavior. The magnetic storms around the corona of the sun are tricky enough. Adding the additional effects of belts of gravitational waves emitted from an approaching black hole, the predictions become indeterminable and colossal. It is now known that when Jupiter, Saturn and Mars line up, shortwave radio frequencies are disturbed and in addition there is a strange and unexplained correlation between this conjunction with violent electrical disturbances in the earth's upper atmosphere.

2012 alignment of the earth behind the sun facing the black hole entrance of the dark rift opening in the Milky Way galaxy.

The ancients associated the "dark rift" or "womb of the Cosmic Mother" of the Milky Way with cataclysmic destruction, death, then the rebirth of a new age. The Mayans today look at the date of 2012 with dread and fear. There is hardly a culture that doesn't identify the end of an age with a cataclysm.

There will be another significant "sign" in the heavens, that will occur in a few months on Saturday, November 8, 2003, that foretells the shift of the age. This event can also be viewed 1:29 a. m. on Sunday November 9, 2003, Greenwich Mean Time (GMT) in England. A total lunar eclipse will occur at 8:12- 8:33 p. m. EST that coincides with a planetary alignment of the Sun, Moon, Jupiter, Saturn, Mars, and Chiron. At the moment of the total eclipse, a Grand Sextile pattern is formed that is not place specific or limited geographically for viewing at any location on Earth. The pattern of the Grand Sextile is known as the Star of David or Solomon's Seal, formed of two inverted grand trines, with the Earth placed in the

513

center and using Chiron as the sixth point. This rare astrological configuration was said to have appeared in the heavens about 5,000 years ago heralding the birth of King Solomon and has not happened since. Can you imagine the awe and beauty of the planet Earth placed inside the balanced symmetry of the six-pointed star during a full lunar eclipse? The Grand Sextile symbolizes the perfect unity of Heaven and Earth or another "sign" of the Union of the Polarity and is the harbinger of some significant event that will soon occur.

Are the authors of this book being too presumptuous to think that it may herald the publication of this book? Is it only by coincidence that this book will be finished by the time the Grand Sextile appears in the heavens; a book that restores knowledge lost for 6,000 years whose very symbol is the Grand Sextile, Star of David or Seal of Solomon? This book, after all, is what the Grand Sextile is all about. We even illustrated the event on the front cover of this book. The restored ancient knowledge of the union of the two earths will forever change how one perceives the creation. It wasn't until July 2003, that we had any knowledge of the approaching Grand Sextile alignment in November 2003, and we were stunned by the timing of completion of this book to the upcoming celestial event. The mystics believe that during this grand celestial event that all the gates of heaven will be open, and the earth will be in a powerful energizing vortex that will "attune" it for the arrival of a new age.

Is Precession a warning of the ancients? Or is Earth merely approaching the Mayan revered end-time as a zero point in time to then restart, once again, time in a new age? Will it be cataclysm or cosmogenesis 2012, or both? The prophecies all speak of The Time of Trial on Earth, Judgment Day, The Time of Great Purification, The End of the Creation, The Quickening, The End of Time as we know it, The Shift of the Ages, The Golden Age, The Millennial Reign, and The Age of Aquarius, which we have now entered, having left the Age of Pisces.

For the past 2,000 years or so mankind has been living in the Age of Pisces, a sign associated with Christianity and spiritual knowledge. We officially entered the Age of Aquarius on March 20, 2000. The dawning of the Age of Aquarius is when the moon is in the Seventh House and Jupiter aligns with Mars. Then shall peace guide the planets, and love will steer the stars. Aquarius is symbolized by the water bearer or the servant of humanity pouring out the water of knowledge to quench the thirst of the world. It represents an age of new knowledge, new ideals, a time of healing, peace, love and spiritual enlightenment. It is a time that has been eagerly awaited for by the world. Will the Precession of the Equinox merely mark a celestial calendar event in 2012, then pass quietly, unnoticed by the majority of the world, and uneventful as many believe that it will? Only time will tell for sure.

As December 21, 2012, approaches, there will be much more information available than currently published to aid and prepare us for this great event. Until that time so be it. **AMEN and AMEN.**

E. J. Clark & B. Alexander Agnew, PhD

End Notes

[1] *Revelation 21:4*

[2] *Isaiah 26*

[3] *2 Peter 3:10*

[4] *Mormon 9:2*

[5] *D&C 101:24, 25*

[6] *Qur'an The Romans 30:50*

[7] *Qur'an The Romans 30:19*

[8] *Qur'an* Fatir *The Originator of Creation 35:9*

[9] *2 Peter* 3:12, 13

[10] *Revelation* 21:1

[11] *Qur'an* Abraham 14:48

[12] *Journal of Discourses* 1X, p. 317; talk given July 13, 1862

[13] (John Taylor, *The Vision* (Salt Lake City, Utah: published by the compiler in 1939; republished by Bookcraft in the 1960's p. 146)

[14] 2 *Corinthians* 12:2

[15] *John* 14:2-3

[16] (*Hymns and Mysteries* in *The Dead Sea Scrolls Uncovered,* Robert Eisenman & Michael Wise, Penguin Books, 1993, p. 228)

[17] *D&C* 98:18

[18] (prophecy as told by Dan Troxell, direct descendant of the Thunderbolts, Cherokee)

[19] (talk given by Credo Mutwa, an 80 year old Zulu Shaman at the Living Lakes Conference in California, Oct. 2, 1999)

[20] www.sec.noaa.gov/weekly/pdf2003/pdf147.pdf *Daily Particle Data* table.

[21] (Graham Hancock, Fingerprints of the Gods, Mandarin, 1995)

The End

Dr. B. Alexander Agnew and I have taken the history of the earth, full round, from its birth to its ultimate destiny. As always, anything new is usually controversial and I would not have it any other way. It is by controversy that things are analyzed and truth becomes evident. It is my fondest hope that the readers will now view the creation from a much grander eternal prospective and not limit their sight solely on the more narrow temporal aspect. By focusing on the more narrow aspect, our vision has become "tunneled", oblivious to the greater spiritual components of the creation. This is where so many have stumbled. Hopefully this stumbling block has now been removed and the Gates of Knowledge re-opened for new interpretation of the creation through the restoration of ancient knowledge referred to as an "old mystery" by the ancients; a "mystery" or "lost truth" that has been hidden for 6,000 years! The restoration of the lost ancient knowledge, an "old mystery", is a great spiritual breakthrough that is symbolized by the Star of David,

Solomon's Seal, or The Grand Sextile pictured on the front cover. Seemingly, it now appears that it was not by chance or coincidence that the manuscript of this book will be finished and the knowledge restored a few days before the Grand Sextile alignment appears in the heavens announcing its return!

Jewish Kabbalists might interpret this alignment as the long a waited sign "son of man" that heralds the imminent return of the Savior. In 1994, when the Shoemaker comet struck Jupiter, they believed this event to be the heralding "sign." Personally, I would think that the Star of David alignment better fits the scenario. Whichever event the "sign" heralds, and it could be both or something else completely unexpected, will be of utmost importance.

However, *The Zohar* foretells the return of the knowledge of the "one mystery" which is the "lost ancient knowledge" that this book restores right before the imminent return of the Savior to renew the world. In other words, this book fulfills that long awaited prophecy. [1] The conditions in the Middle East seemingly point toward "ripe" conditions for the return of the Lord but bear in mind that "imminent return" could mean nine years into the future which is the year 2012! I know that you are simply dying to know what The Zohar prophecy states, right? It states:

"And observe that all these measurements prescribed
for this world had for their object the establishment
of this world after the pattern of the upper world, so
that the two should be knit together into one mystery

(the lost truth). At the destined time, when the Holy One, blessed be He, will bestir Himself to renew the world, (the second coming) all the world will be found to express one mystery (come into knowledge of the lost truth), and the glory of the Almighty will then be over all, in fulfillment of the verse, "…in that day shall there be one Lord , and his name one."[2] It is widely accepted by latter-day scholars that whenever prophets use the term *in that day* they are referring our day; right now.

Speaking for myself, once my hand was set to the plow, in writing this book, there was no turning back. It was as if I were driven by unseen forces to complete the task; driven to the point of getting up in the middle of the night to make corrections and additions to chapters. For two years it consumed my thoughts to the point of obsession, yet, I have loved every word that was written, therefore this work is a "labor of love." I now truly understand the meaning of that statement. Those who were spiritually awake were drawn to me, wanting to hear of the progress of the book, new finds, etc. and those who were spiritually asleep actually avoided an encounter whenever possible, not wanting to hear "boring" talk about the endeavor. As some put it, "its too deep for me."

Regardless of the matter, B. Agnew and I tackled writing on a very difficult subject, to say the least, a subject most authors wouldn't touch with a ten-foot pole. The only reason that I considered writing the story was that it was too important not to. I chose Dr. Agnew to co-author because he and I think along the same lines plus he is receptive to new ideas and fearless in un-chartered waters. Once the plot was laid, he embellished it. I gave the meat and he gave the potatoes—or steak and sizzle for those of you who only eat meat—so to speak. We sort of complement each other's writing.

Quite unexpectedly, I was moved spiritually when I envisioned the great "stacked up balloon universes" like galaxies in the vastness of unending outer space. Through this inspiration the sources of evidence confirming this arrangement began flowing like a wellspring. I don't know what astronomers will discover, but the following illustration is a close rendition of that vision.

Multiple stacked universes as illustrated by E. J. Clark

It was a humbling experience to realize that the creation is much more than our universe and the immensity is far beyond our mortal comprehension. I felt infinitely insignificant, unimportant and small in comparison and yet at the same time "honored" to have been permitted to envision them.

Another realization that I became more aware of is that everything has a spirit and a life, including plants, animals, and the earth. I'm now more careful with the environment, pausing more to enjoy its beauty, and respectful of the sanctity of life. Recently I saw a turtle trying to cross a highway and taking pity on the creature, I pulled my car off the road, picked it up and carried it to safety on the other side. I had a good feeling the rest of the day, kind of like doing a good deed for someone. Life has value whether it is an ant, tree or pet. All creatures should be treated with respect and kindness, as the Earth would be a dreary place without them. Each serves its assigned role or else the planet would be barren and empty. As Adam was put in the Garden to dress and keep it, we have inherited his job as caretakers and overseers of this creation. As responsible stewards we are not only to guard, preserve, and protect nature but additionally expected to conserve and make it even more beautiful and fruitful than when we received it.

My soul feels satisfied that most of the questions are answered reasonably well concerning our earth, questions that have perplexed scientist and religious leaders since the loss of the ancient knowledge, now restored. The jigsaw puzzle is now complete.

Once more my daughter Eve, asked, " Can you prove that such an event actually occurred?" I replied, "Does it answer all your questions?" She slowly said, "yes." "Then it must have happened" was my reply.

Since the ancient "lost truth" is spiritual in nature it cannot be proved as such, but it does answer all the questions. The proof here is truly in "the pudding." I perceive that the ancients lost this knowledge due to unbelief, and lack of understanding. Their loss of faith in this ancient truth has hampered us for 6,000 years, causing many to stumble and fall away in disbelief. Sometimes proof requires us to take "that giant leap of faith" without questioning until the times of understanding are revealed. That time is now! We have arrived to a new level of understanding and awareness of the creation both spiritually and temporally. And the best part is it is only the beginning of what is to come in knowledge concerning our planet and its ancient history. The restoration of the "lost truth" is the key that will re-open the Gate of Knowledge to our best "thinkers."

I have been so elated over being a part of the restoration of the "hidden mystery" that my soul has literally been singing with joy. Does that sound crazy? Have you ever had a tune that has stuck in your head all day causing you to hum along? Well, that is your soul singing. Unfortunately, my soul has been singing the same tune for nearly two weeks, *Morning Has Broken,* and it is driving me nuts. The irony here is that it is a song on the creation.

My greatest aspiration over the next few years is to revisit the Museum of Anthropology in Mexico City, the Cairo Museum in Egypt, and take a first time tour of the British Museum in London. Machu Picchu is also a top priority destination as I want to touch my forehead to the forehead of the Intiwatana; it is said to impart enlightenment to the third eye of wisdom. Sound like strange requests? Not at all, I only desire to see and experience first hand what I have written about as opposed to looking at pictures.

Well, I guess that about winds things up. Take care and God Bless.

E.J. Clark

Truth is truth. In my first writings many years ago I stated, "Everything we do will eventually affect the universe." I still believe that. Yes, people place good and evil effects into motion. But, perhaps the saddest loss is from they who do nothing. They think nothing. They have no imagination. They consume and are consumed in the machines of mortality. I suppose that even this has its effect on the universe—like fodder for the engines of kings and presidents—but the act of reading this book is the result of my action upon the universe.

Your viewpoint about the creation of the earth has been changed in some way by reading this book. Words you may have read hundreds of times before now have new meaning. Fleeting wonder about the works beneath your feet and in your lungs and coursing through your veins will blossom with new intensity and color to the limits of your imagination, as you reflect and ponder on what you see and hear from this day forward.

This book may be one of the greatest landmark works in the genre, because it relates man's knowledge of the creation with a restored viewpoint. It was prophesied millennia ago that in the latter times the knowledge of how this earth came to be, and where it is destined to go, would be restored. There have been numerous punctuating events that needed to happen first, the most recent of which was the Grand Sextile formation with a full lunar eclipse— see the cover artwork. The last time this occurred, King Solomon was born. Now it is time for this book to be, and for you to stand upon it and reach even further into the heavens, toward God if you can.

We want you to pick up this book again and again throughout your life and perhaps learn even more, that you may not have been prepared to learn this time around. We want you to tell your friends, use it in your Sunday school classes, and draw from your feelings during those campfire discussions while looking up at the stars. We hope that you have shared in our enthusiasm and occasional ecstasy as we discovered long-lost knowledge and heard the voices of ancient writers who sealed their records with their very lives.

Maybe you'll feel the rumble of the earth as continents sunk beneath the sea while heavenly bodies threatened complete destruction of the human race. Maybe you'll be able to imagine how the sky over Atlantis or Mu appeared full of flying machines. Maybe you'll squint from the gleam of the sun off crystal buildings or the gold laminations of the temples. And maybe, just maybe, you will question modern man's rhetoric of how the fact the earth spins on its axis was only discovered in the 1850's or how we have advanced further in the last 100 years than in the last ten millennia man has been on the planet.

Now you know that there have been many civilizations, some with more advanced technology than our own, wielding their dominion on the earth to their destruction. Their edifices have been misinterpreted. Their glyphs have been misread. Their records have been destroyed out of superstition or by the evil machines of wicked men whose greatest weapon was maintaining complete ignorance of the masses.

Well, this book changes that forever. Your having read it changes that forever. Now that you know there are no limits to your potential, what are you going to do with it?

—B. Alexander Agnew

End Notes

[1] *(The Zohar, translated by Simon, Sperling and Levertoff, ©1984, London, The Soncino Press, Ltd., Vol. 1V p. 299)*

[2] *(Zechariah* 14:9) Parenthesis and bold type added.

Index

Australopithecus 400. *See* Lucy;
   See Lucy
Autogenes 43
Avims 335
Azkeel 167

B

Ba'albek 351
Babylonian 3
Babylonian Talmud 2
Balaam 155, 294, 296, 297,
   298
Balam-Quitze 361
Barbelo 43
Batraal 167
Battle of Allia 347
Berosus 3
Bhagavad Gita 12, 506
Bible xiii, 2, 4, 7, 8, 10, 21, 35,
   45, 46, 53, 61, 67, 108,
   135, 175, 226, 227, 233,
   285, 298, 303, 307, 325,
   333, 335, 337, 350, 351,
   352, 353, 374, 465, 473,
   482
Big Bang 110, 111
Black holes 17, 27, 28, 49
black holes 27, 28, 31, 114, 135
Boaz 261
Book of Abraham 3, 132, 291,
   292, 326
Book of Enoch. See Book of
   The Secrets of Enoch 3,
   111, 154, 162, 167, 168,
   193, 194, 212, 238, 306,
   307, 467, 489
Book of Jasher 4, 173, 175,
   181, 188, 193, 194, 213,
289, 293, 294, 296, 298,
   305, 326, 418, 476.
   See Book of the Ancient
   World and The Book of
   the Upright; *See* Book of
   the Ancient World and
   The Book of the Upright
Book of Mormon 4, 67, 285,
   361, 363, 364, 365, 366,
   376, 478, 481, 482
Book of Moses 5, 54, 175, 302,
   303
Book of the Dead.
   See Kabbalah; *See*
   Kabbalah
Book of the Great Awakening 6
Brigham Young 65, 68, 83,
   215, 496
brother of Jared 482.
   See Mhanri Moriancumer;
   See Mhanri Moriancumer
Buddhism 42, 61, 495

C

Calius Julius Maximinus 348
Cambrian 117, 119
Canaanites 335
Caphtorims 335
Cayce 321, 322, 323, 327, 440
celestial 36, 71, 83, 133, 161,
   178, 202, 214, 220, 222,
   223, 224, 228, 231, 445,
   455, 497, 498, 506, 514,
   515
Celtic 254, 270, 343, 346, 347
Celtic cross 254
Celts 343, 345, 346, 347, 348,
   350

three degrees of glory 42
Throne of Glory 38, 39.
    See Seven heavens; *See*
    Seven heavens
Tiahuanako 273
Tla-acoma 481
Tla-acomah 481
Toltec 361, 363, 478, 481, 482.
    See Nephite; *See* Nephite
Toltecas 363, 366. *See* Toltecs;
    See Toltecs
Toltecs 363, 479
Tonatiuhican 42
Torah 2, 11, 13, 290, 492
Tower of Babel 323, 474, 480
Tree of Knowledge 154, 159
Tree of Life 154, 158, 162,
    163, 505
Turel 167

U

UFO 434, 504
Ulmecas 363, 364
uniformitarianism 217
union of the polarity 32, 234,
    241, 243, 246, 247, 248,
    252, 254, 255, 259, 260,
    261, 262, 263, 265, 268,
    269, 272, 273, 274, 277,
    279, 283
Universe 39, 60, 68, 127, 141,
    426, 429, 495, 496
Urakabarameel 167
Urim and Thummin 82, 83
Utnapishtim 2

V

Vailxi 424, 425

Vedic 11, 61, 132, 292, 325,
    424, 506
Venus 198, 202, 205, 206, 217,
    222, 223, 438, 499, 500
Vimanas 424, 432. *See* Vailxi;
    See Vailxi
Vortex 22, 433
Votan 323, 478, 479

W

Wakea 480. *See* The Fair God;
    See The Fair God
Washington 19, 257, 259, 260,
    261, 311, 313
Washoe 340
Watchers 169, 214, 458
watchers 69, 166, 167, 329,
    396, 403, 404
Watusi 370
Wicca 299
Wilkinson Microwave
    Anisotropy Probe 110
William St. Clair 318, 319
WMAP 110, 111. *See* Wilkinson
    Microwave Anisotropy
    Probe; *See* Wilkinson
    Microwave Anisotropy
    Probe
World of Assiyah 39
World of Atsiluth or Emanation
    39
World of Beri'ah or Creation
    39
World of Yetsirah of Formation
    39

X

x-ray 32

Xisitros 2

Y

Yin and Yang 113, 243, 263,
    264, 274, 282
Yoga 281
Yomyael 167
York Rite 309

Z

Zamzummims 336
Zavebe 167
Zebbaj 198
Zebusites 336
Zeus 412, 413, 417, 439, 447,
    450, 463. *See* Jupiter; *See*
    Jupiter
Zevul 38. *See* seven heavens;
    See seven heavens
Ziusudra 2
Zohar 2, 13, 44, 47, 65, 68,
    70, 91, 97, 108, 155, 161,
    166, 175, 183, 187, 190,
    193, 194, 214, 224, 225,
    228, 234, 238, 243, 283,
    285, 291, 315, 338, 376,
    391, 395, 402, 436, 458,
    466, 475, 476, 489, 492,
    497, 519, 528. *See* The
    Book of Splendor; *See*
    The Book of Splendor
Zoroaster. *See* Zarathustra; *See*
    Zarathustra
Zostrianos 43, 46, 51, 52, 64,
    67, 68, 90
Zuzims 337

About The Author

E. J. Clark is the classic American explorer. E. J. has personally crawled through the passage ways of pyramids, temples, jungles, and museums of ancient history. Every point of interest recorded by ancient interpreters of the creation and metamorphosis of the Earth has been sifted through curious hands for the reader's enlightenment. For 30 years E. J. has been compiling data, pictures, and testimonies for the writing of this book. To say an entire lifetime of research has gone into it would be an understatement. With the passion of an armchair archeologist and the tenacity of an astronomer E. J. has acquired a skill set that uniquely explains the mysteries of the eternities. Dr. B. Alexander is a published PhD physicist with decades of research in the spectrometry of creation. As an exceptional writer and accomplished scriptural scholar he has been able to build the bridge between evolution and creationism. The reader may search for a lifetime and not find a better partnership of authors to reveal the most incredible story in the universe.